BOOKS THAT MAKE YO

"What is a civil engineer's favorite card game?"
ANSWER: Bridge

-Albert B. Squid

SQUARE ROOT OF SQUID PUBLISHING

BOOK CONCEPT BY: ALBERT B. SQUID
ILLUSTRATIONS/WRITING BY: ALBERT B. SQUID

10 ENGINEERING PROJECTS

IN 5 DIFFERENT ENGINEERING FIELDS!!!

CIVIL ENGINEERING

EXPERIMENT 1

TESTING CANTILEVERS

EXPERIMENT 2

BUILDING & TESTING A BRIDGE

AEROSPACE ENGINEERING

EXPERIMENT 1

TESTING AIR PRESSURE ON AIRPLANE WINGS

EXPERIMENT 2

TESTING WING DESIGN WITH PAPER PLANES

EXPERIMENT 2

BUILD A GEAR BOX

MECHANICAL ENGINEERING

EXPERIMENT 1

BUILD A GEAR TRAIN

0011010011100111 0

SOFTWARE ENGINEERING

PRINT("SQUID")

EXPERIMENT 1

SOLVE THE BINARY CODE

EXPERIMENT 2

WRITE CODE

EXPERIMENT 1

BUILD A WATER COLLECTOR

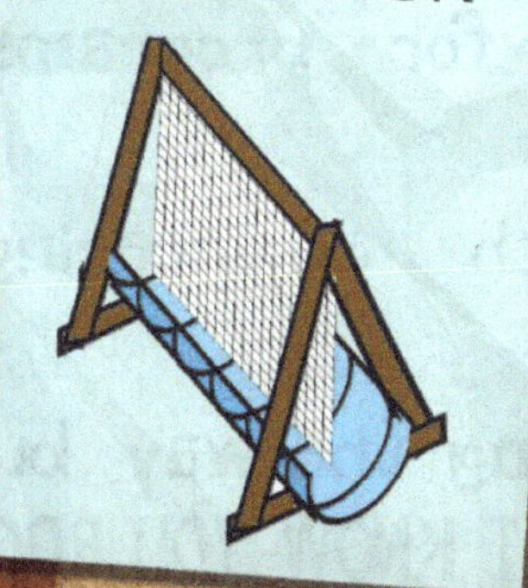

ENVIRONMENTAL ENGINEERING

EXPERIMENT 2

BUILD A COMPOST BIN

HEY THERE ENGINEER!

I'm Albert B. Squid (Eclectic Engineer, Architecture Adventurer, & Wacky Writer) and I was hoping you would help me with some of my engineering experiments throughout this book. The experiments are divided into five branches of engineering: Civil, Aerospace, Mechanical, Software, and Environmental. In each of these branches, we will do some fun activities to find the answers to some burning questions like:

1. How do bridges and buildings hold so much weight without breaking?

2. How do airplane wings work?

3. How do gears work in machines?

4. How does one write computer code for programs?

5. How can we save the Earth with engineering?

Oh! There might be some math (or maths) along the way, but do not fret because you are an ENGINEER... SO I KNOW YOU ROCK IN THAT DEPARTMENT!!!

ENGINEERING

Designing Building and Bridge Structures

EXPERIMENT 1
THE CANTILEVER

What we need!

coins

stack of books

What is a CANTILEVER?

A diving board is a cantilever. The green base is the support. The bigger & stronger the support the further the cantilever can stretch.

Cantilevers are used in a lot of buildings too.

CANTILEVER BEAMS

Let's make four different types of cantilever beams and then test them.

C BEAM

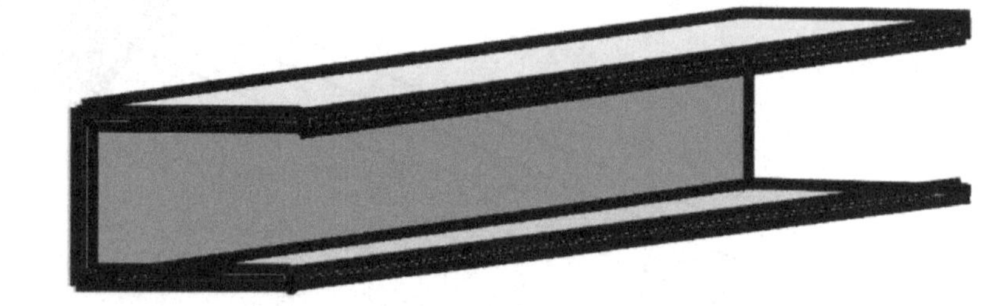

BOX BEAM

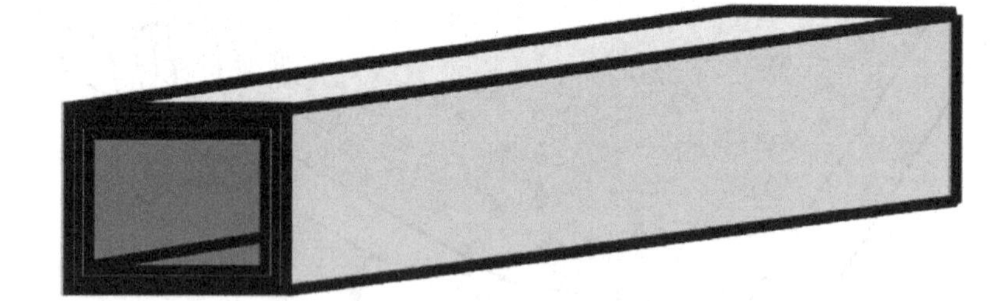

I BEAM

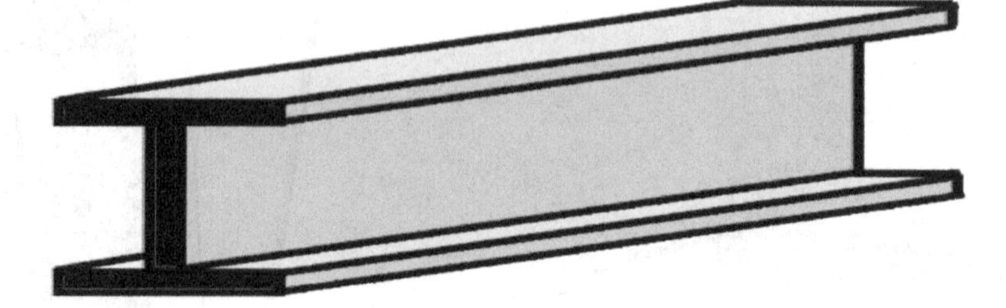

NEW BEAM

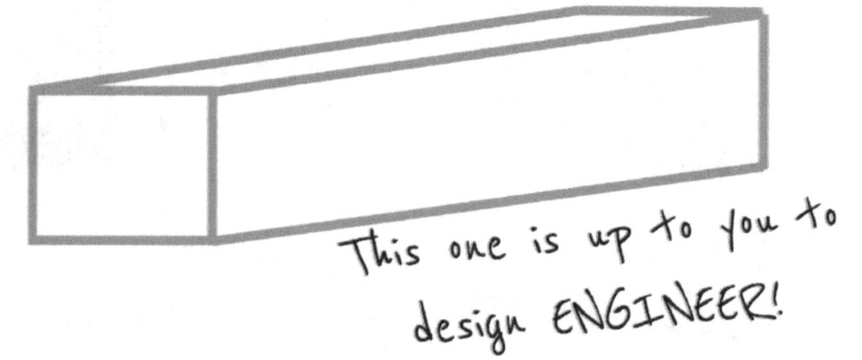

Cut out the two steel (paper) C BEAMS. Fold along the dotted lines to make a "C" shape.

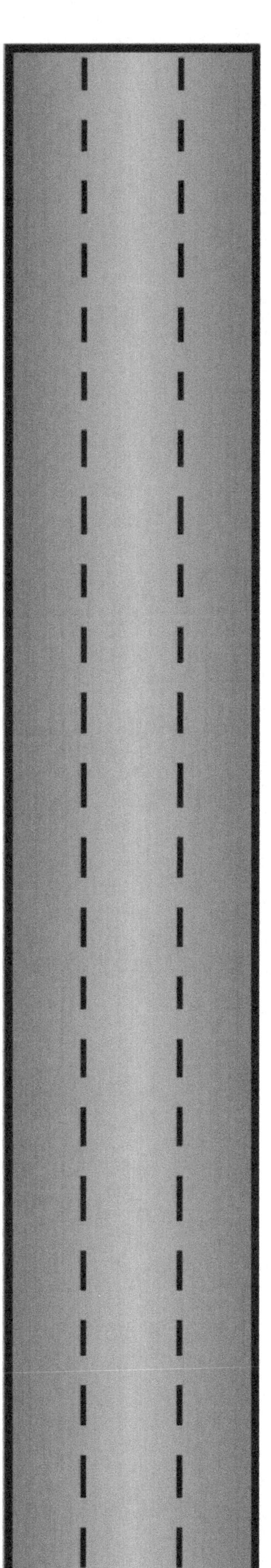

Cut out the two **BOX BEAMS**, *fold* & glue them into a box shape.

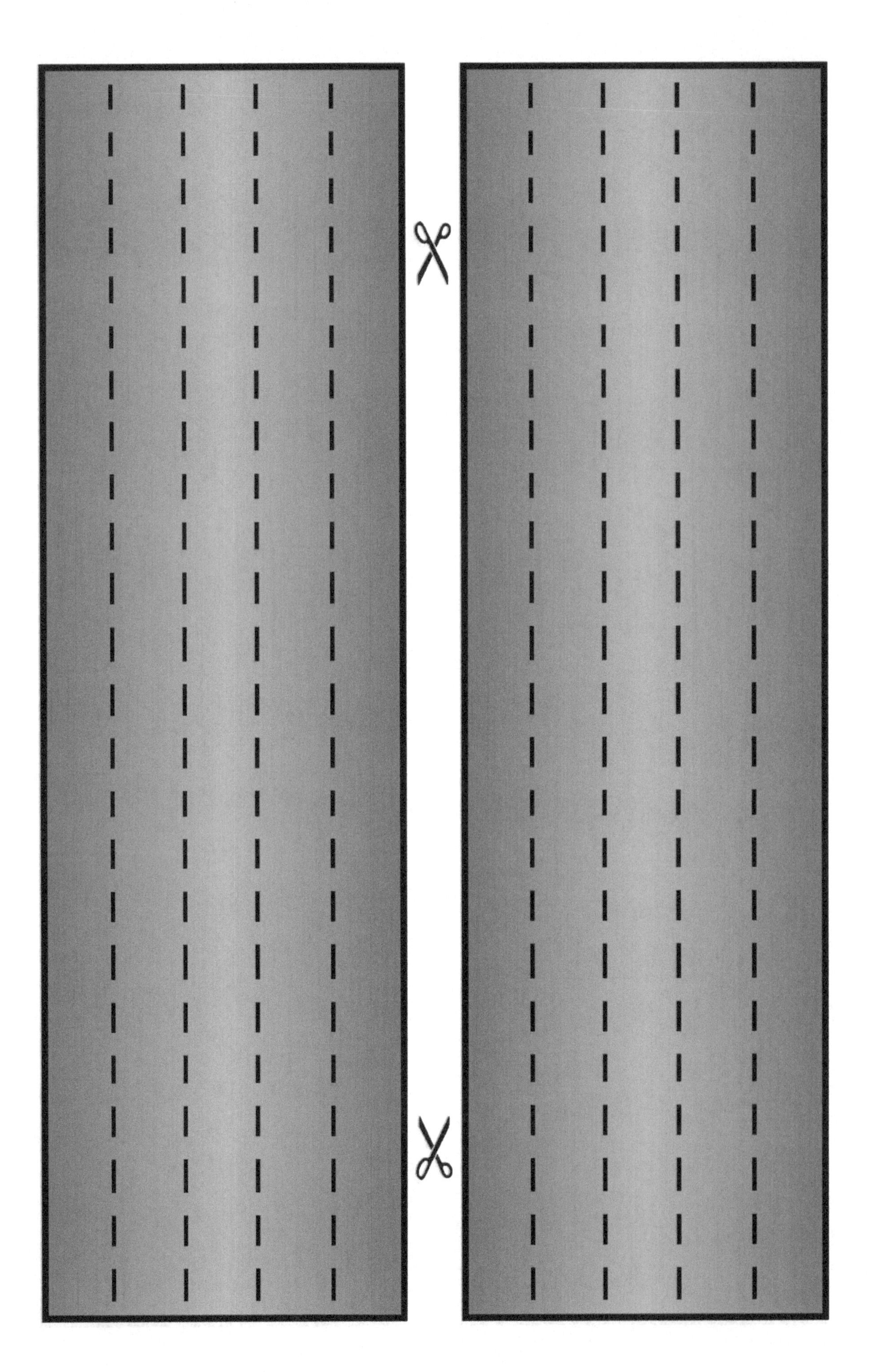

Cut out the I BEAMS. Follow the instructions about how to glue them together.

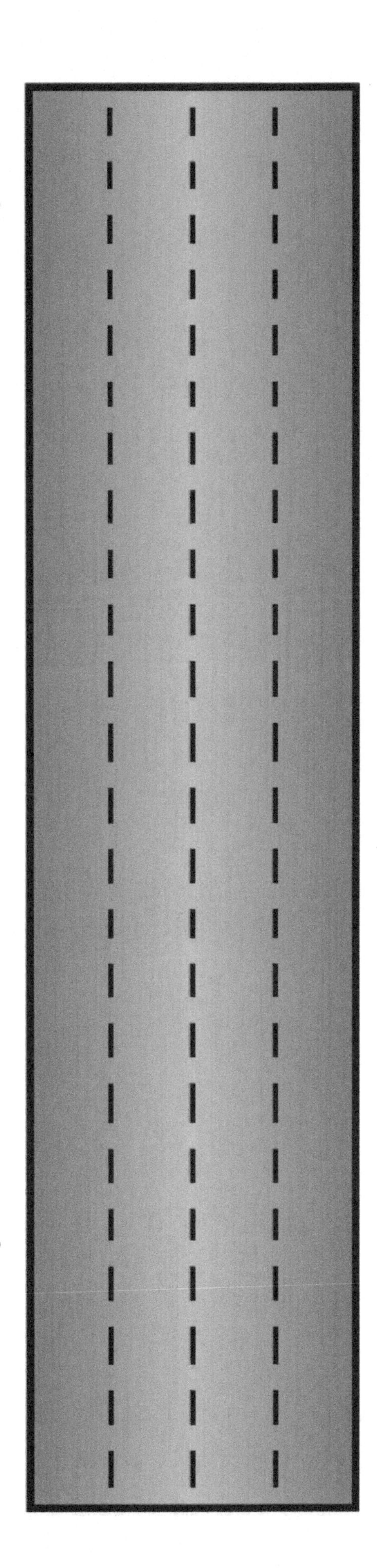

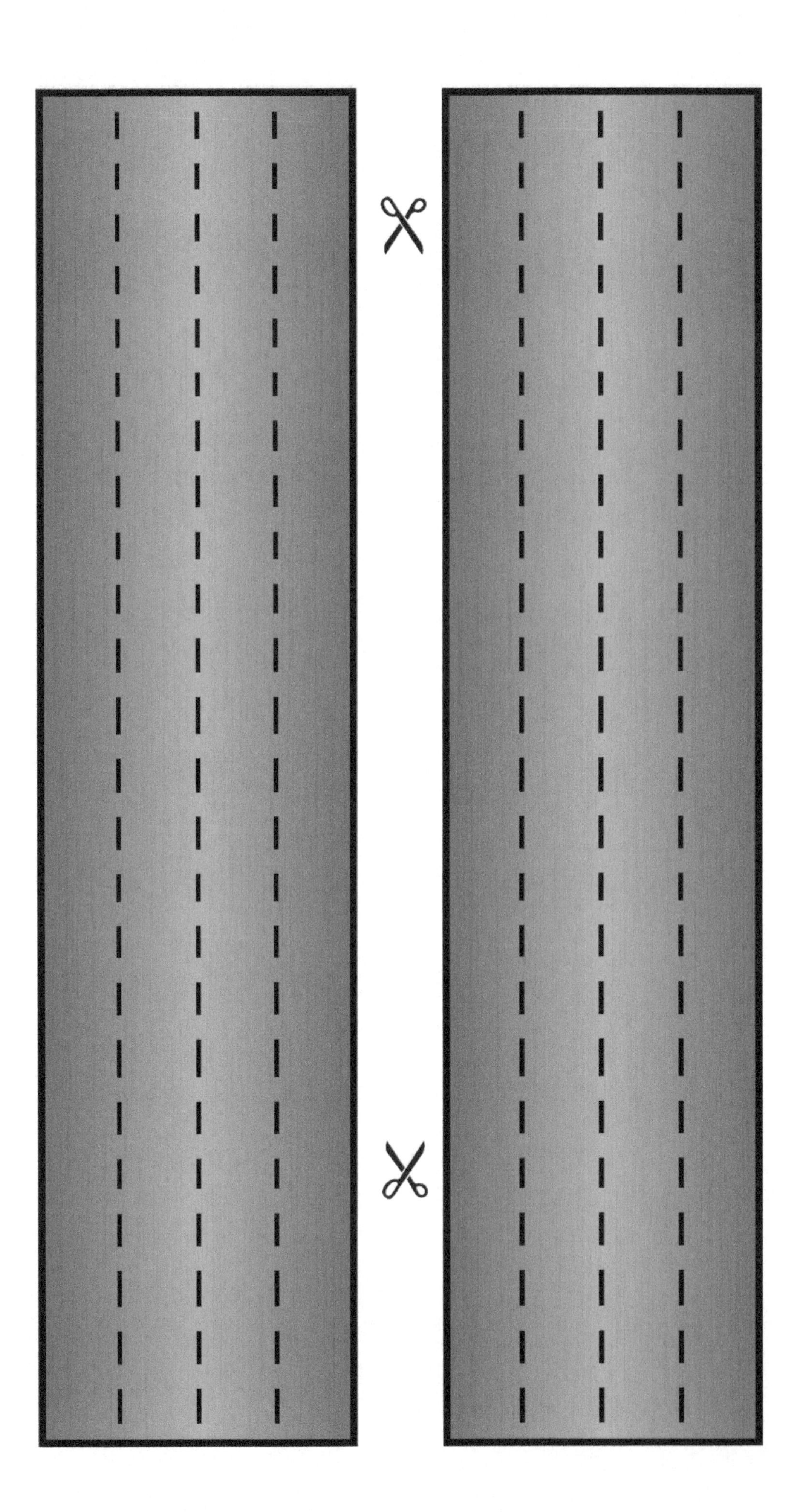

Cut out the I BEAMS. Follow the instructions about how to glue them together.

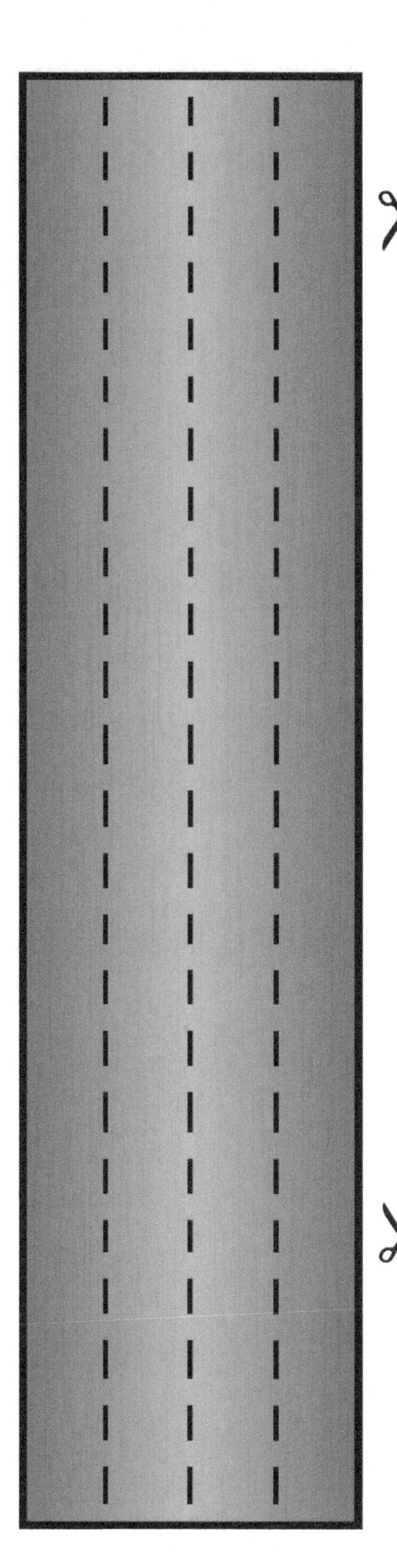

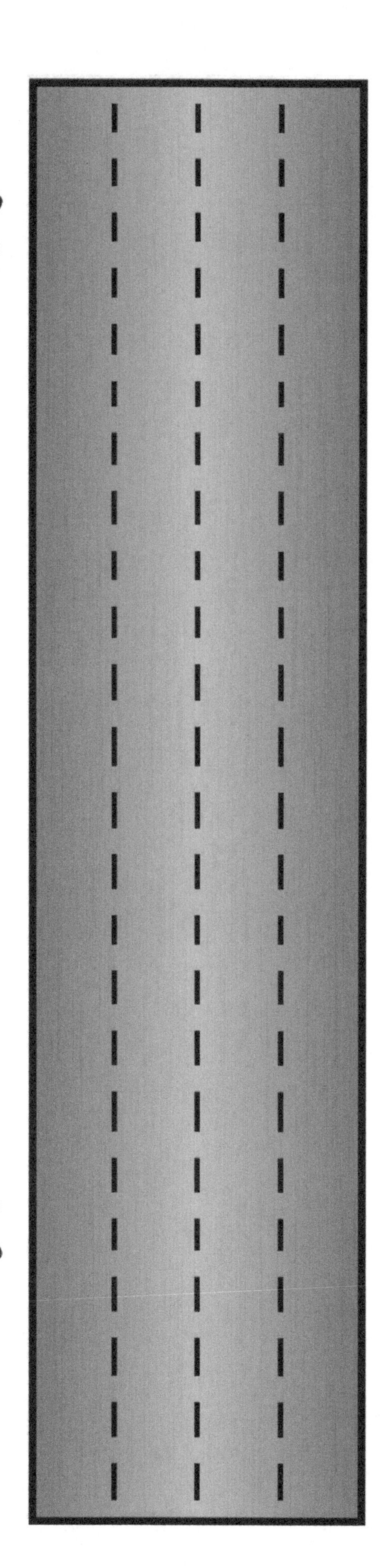

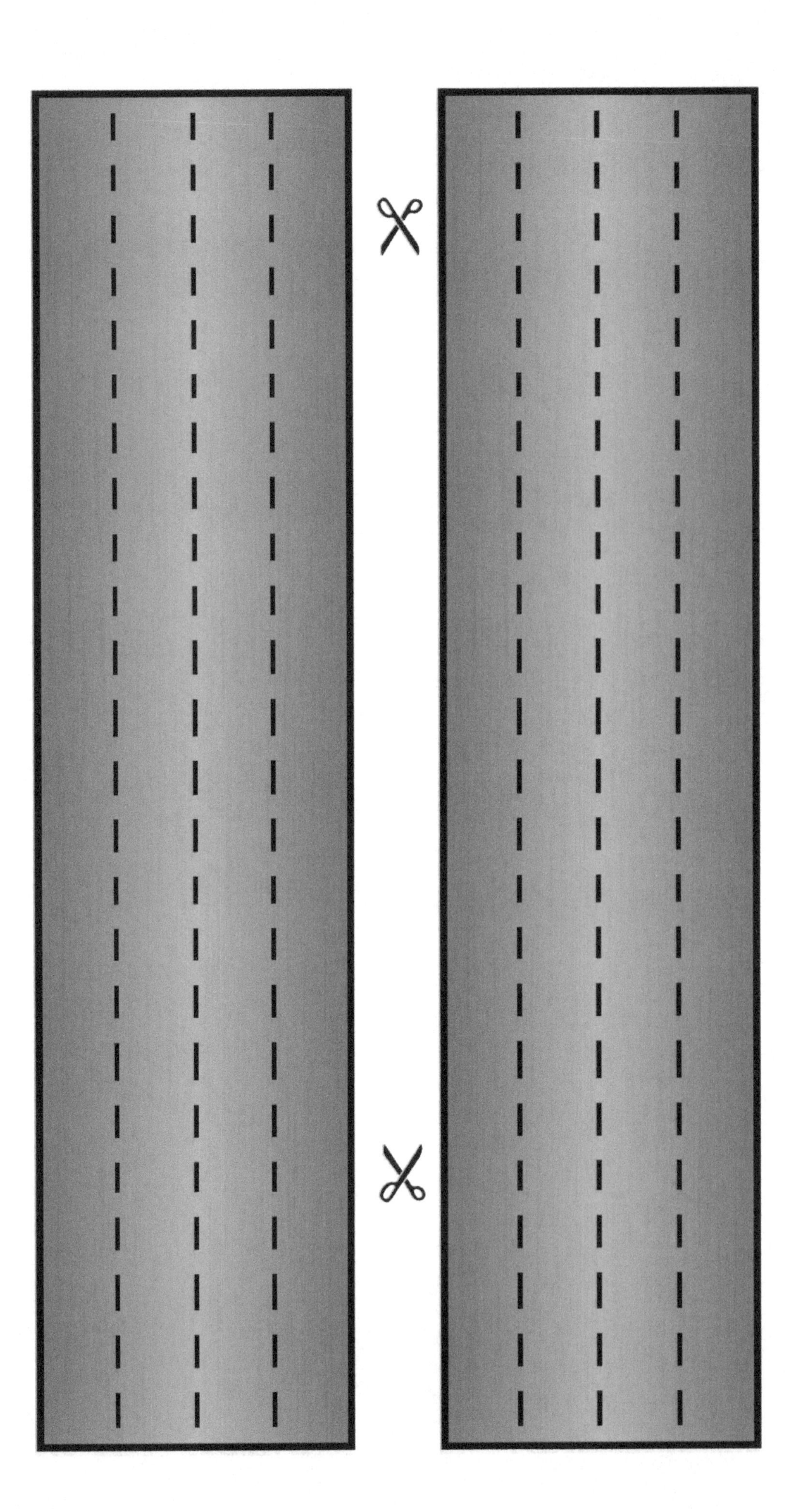

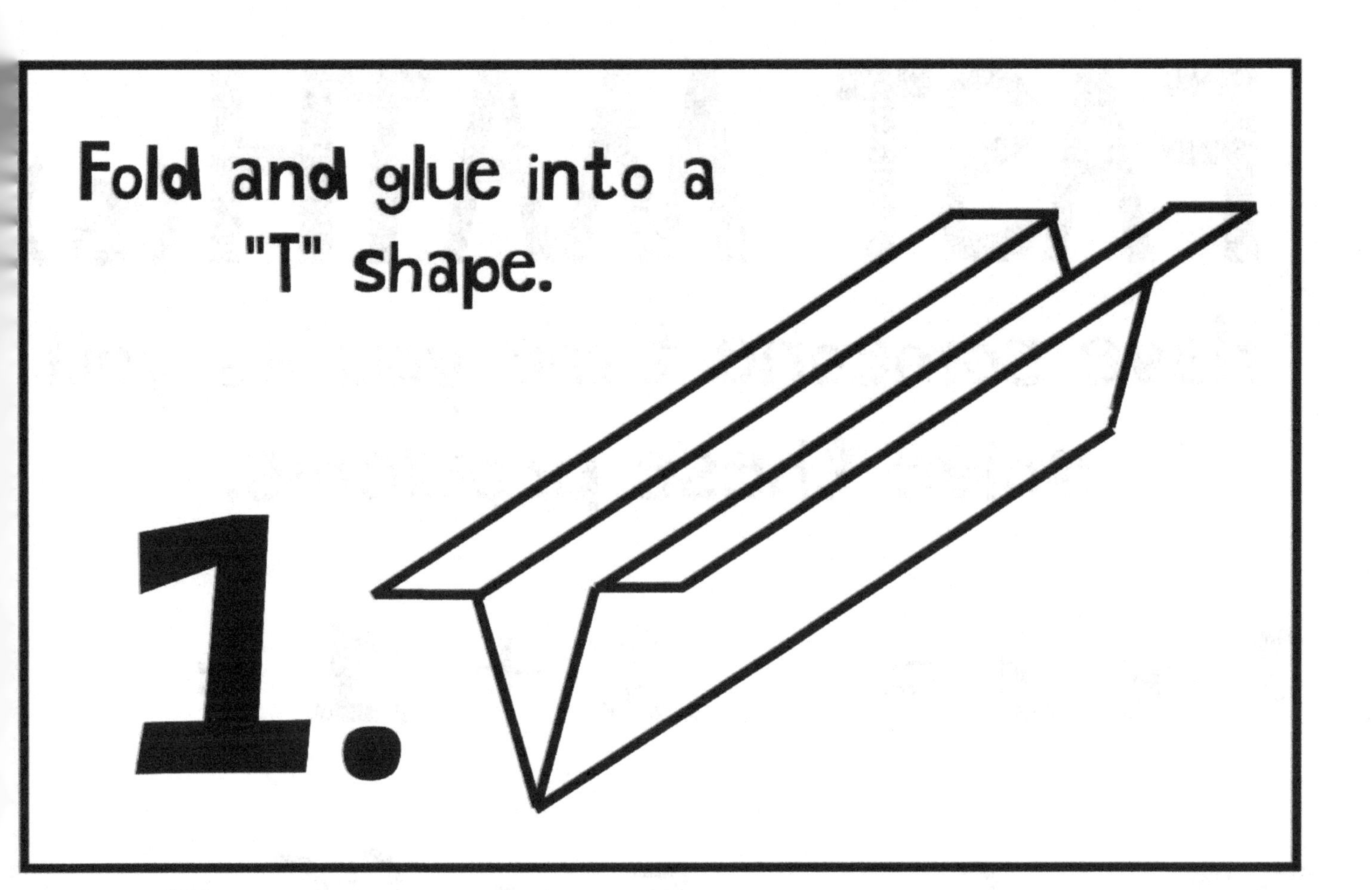
Fold and glue into a
"T" shape.
1.

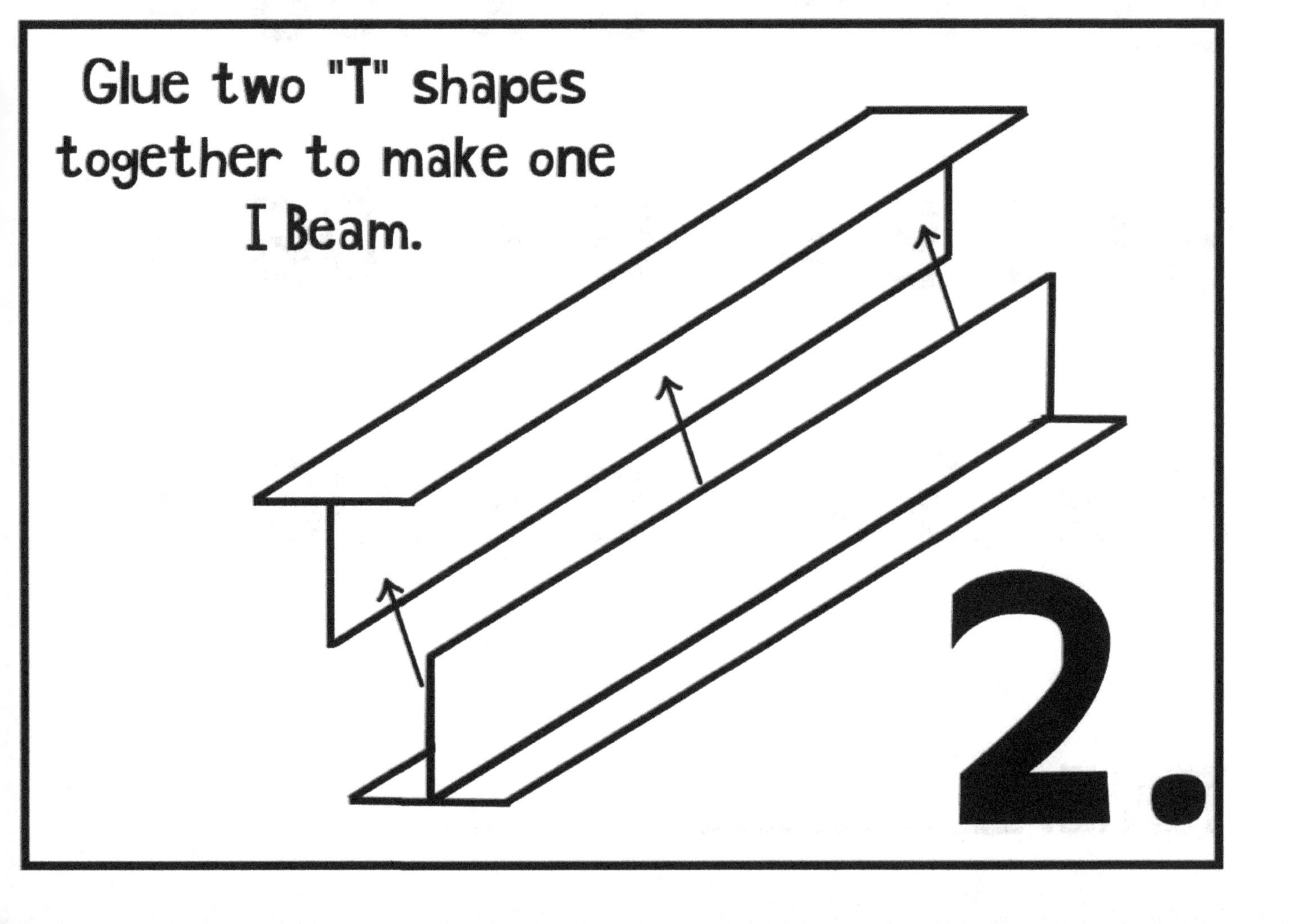
Glue two "T" shapes
together to make one
I Beam.
2.

FAST MATH(s)

Have someone time you as you solve these problems.

$7 \times 9 =$

$.1 + .1 =$

$39 - 6 =$

$5 \times .06 =$

$81 \div 9 =$

$5 + .06 =$

$8 \times .3 =$

$5 - .06 =$

TIME: ______________

This is where you can design your own **BEAM** shape **ENGINEER**. Give it a try and test it compared to the other shaped **BEAMS**.

This is where you can design your own **BEAM** shape **ENGINEER**. Give it a try and test it compared to the other shaped **BEAMS**.

THE TEST

For each beam shape push down firmly on one end on top of the book stack (or table top). Place about ten coins on the other end.

Start with the beam about an inch (2.5 cm) hanging over the edge and slowly push it forward until it breaks. Record the data. Measure where each beam broke.

BEAM TEST RESULTS

BEAM TYPE		NUMBER OF COINS	BREAKING POINT inches/cm	BEAM RATING 1 poor - 10 best
C BEAM	1			
	2			
BOX BEAM	1			
	2			
I BEAM	1			
	2			
? BEAM	1			
	2			

Which beam type did the best? Why?

EXPERIMENT 2
BUILD A BEAM BRIDGE

What is a beam bridge?

A beam bridge is the most simple of all the bridges. It is just a horizontal thingy with two vertical thingies holdin' it up on both sides. These kind of bridges are used for things that are not that heavy.

Most modern beam bridges are made of steel beams, but in our case, paper will work.

BEAM BRIDGE

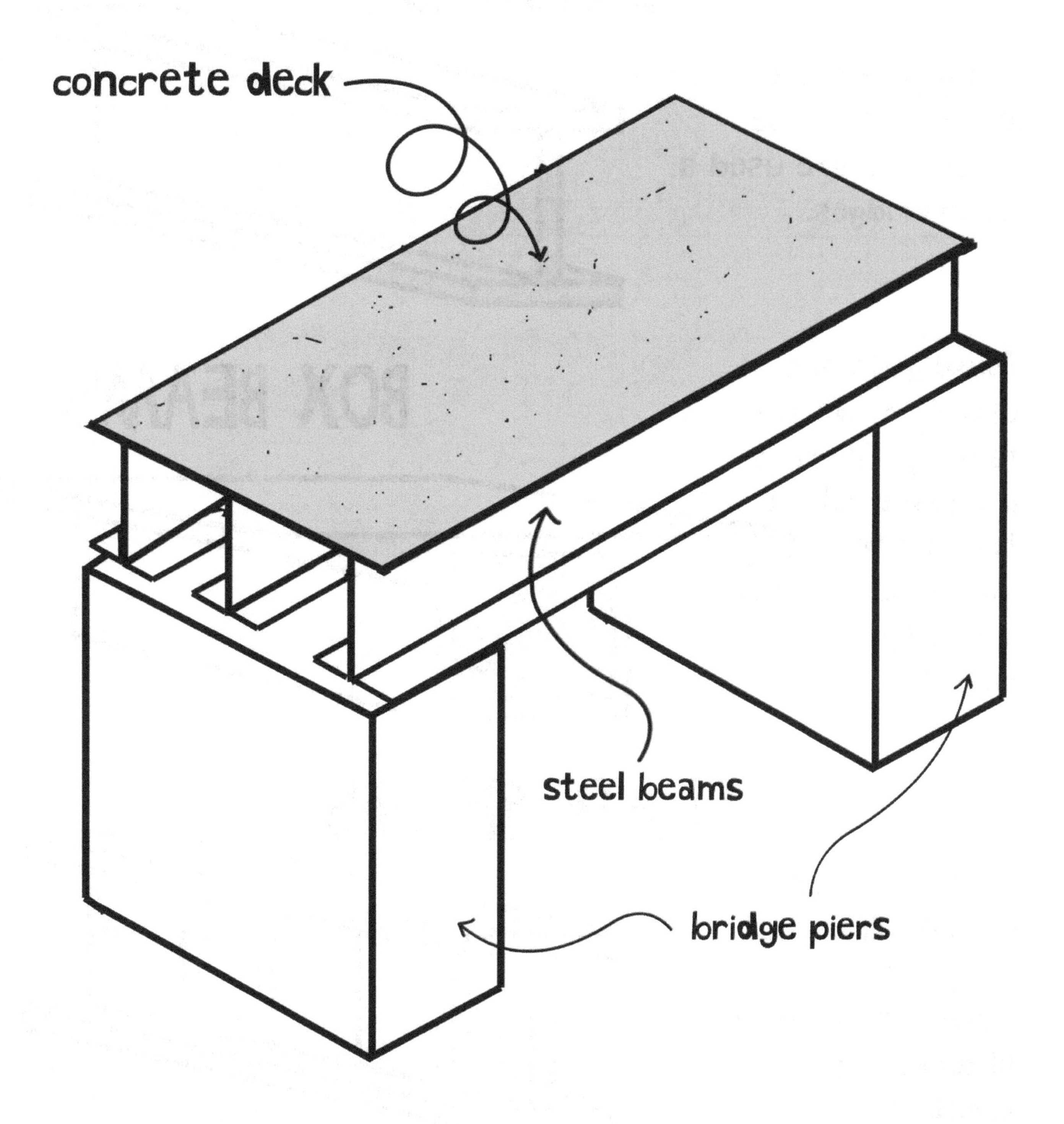

(piled up books will make great piers for our paper beam bridge.)

TYPES OF BEAMS

I BEAMS

Can you see the letter "I"? These steel guys are used a lot in bridges.

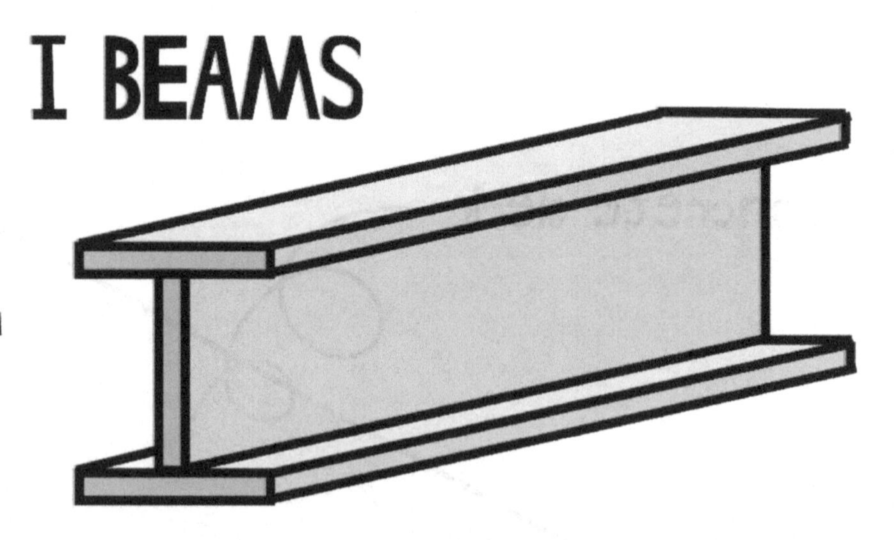

BOX BEAMS

These guys are usually made in a factory. There is a square hole in the middle. Makes sense they are called box beams.

PLATE BEAMS

These are cousins of the I Beam but they are made by welding a bunch of steel plates together.

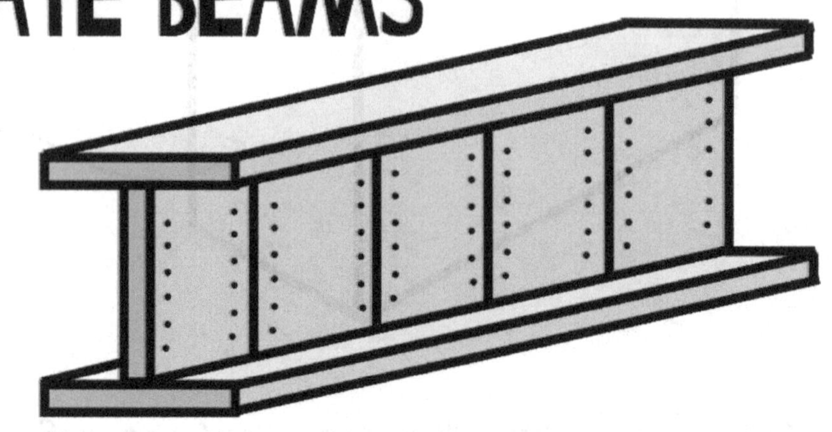

Fold and glue into a "T" shape.
1.

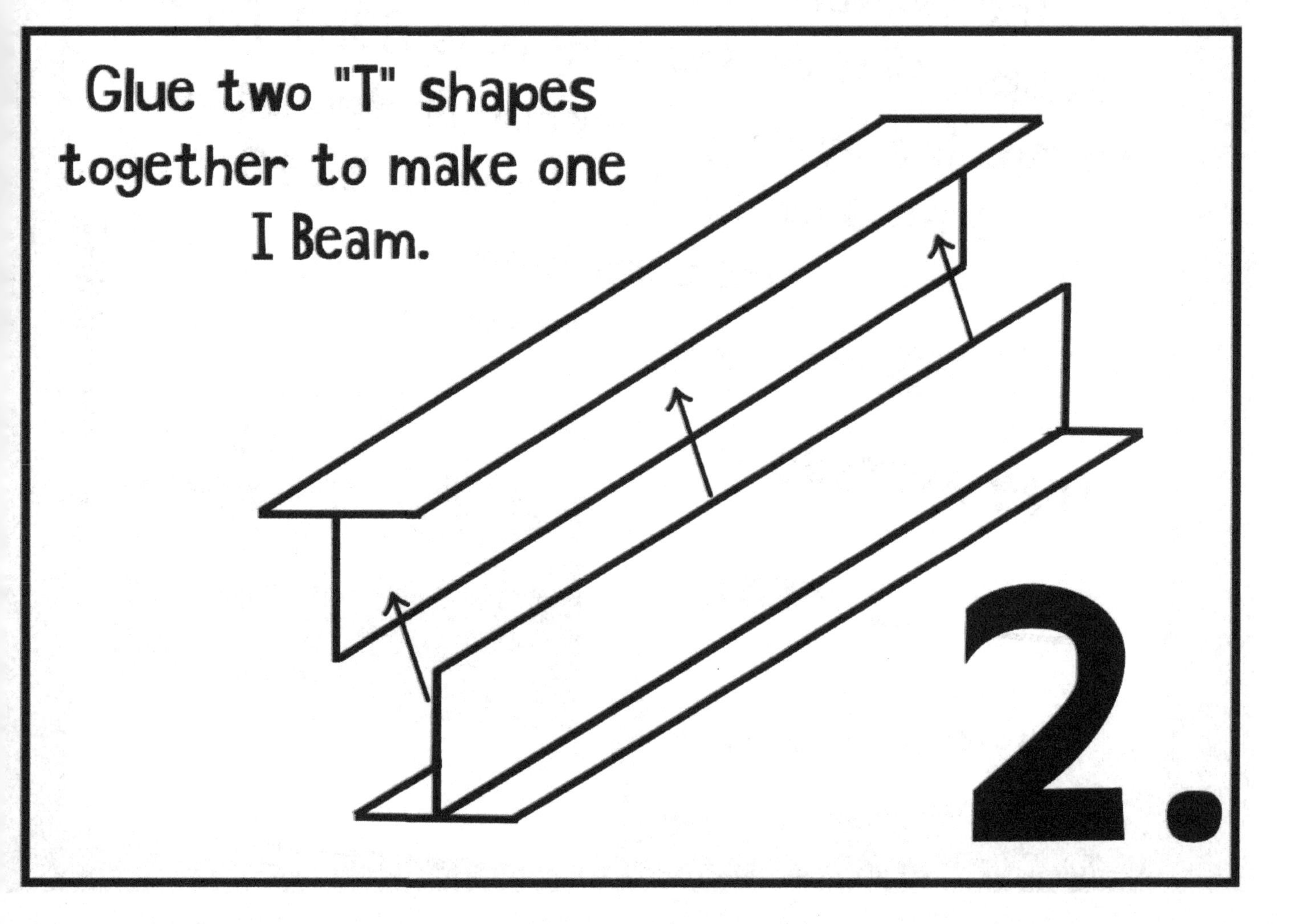
Glue two "T" shapes together to make one I Beam.
2.

CHANGE FEET TO METERS AND METERS TO FEET

8981 FEET =

________ METERS

1 foot = .3048 meters

The Golden Gate Bridge above is a suspension bridge and it is 8981 feet long. The Sydney Harbour Bridge below is an arch bridge that is 1149 meters long.

1149 METERS =

________ FEET

1 meter = 3.28 feet

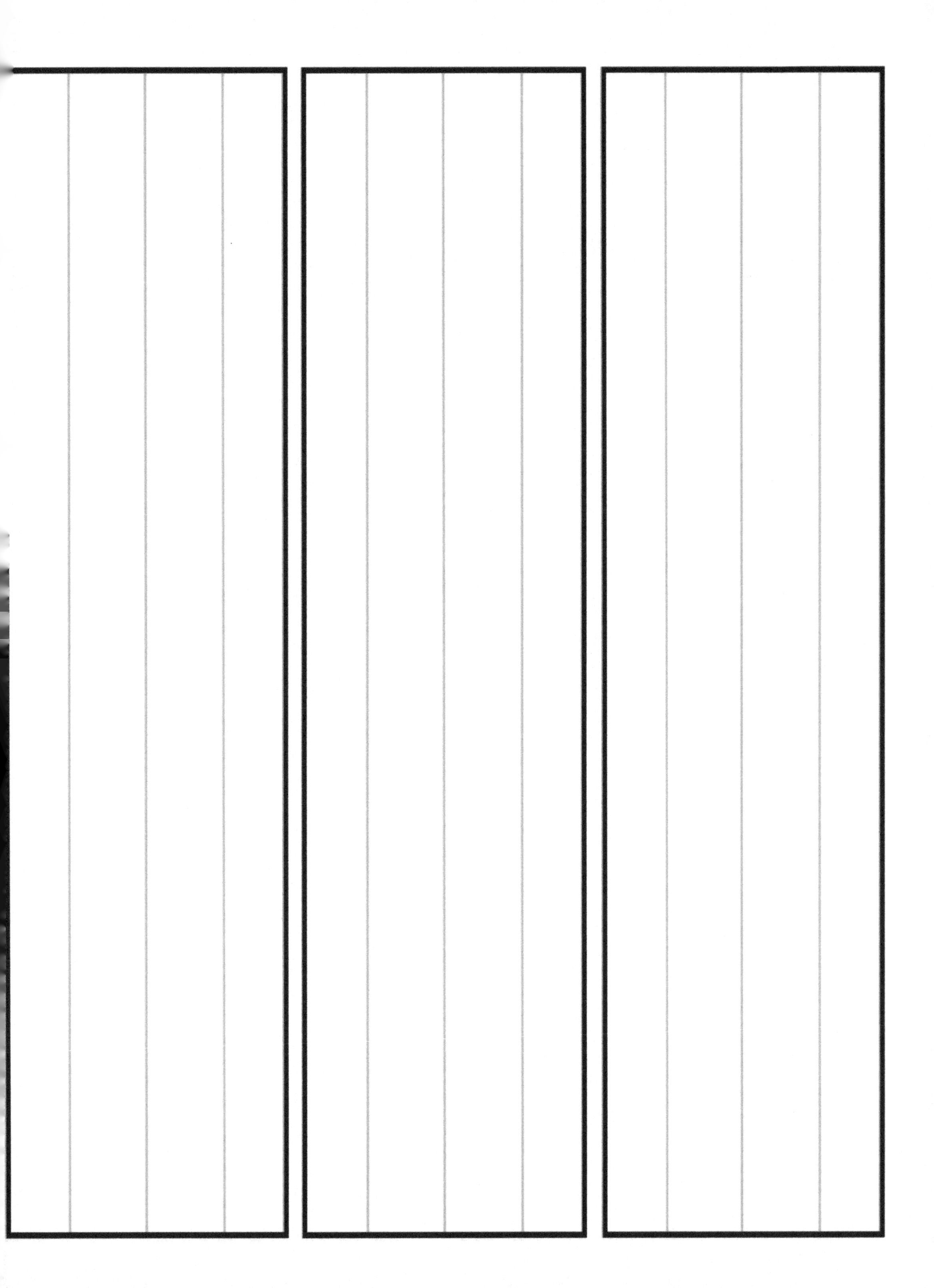

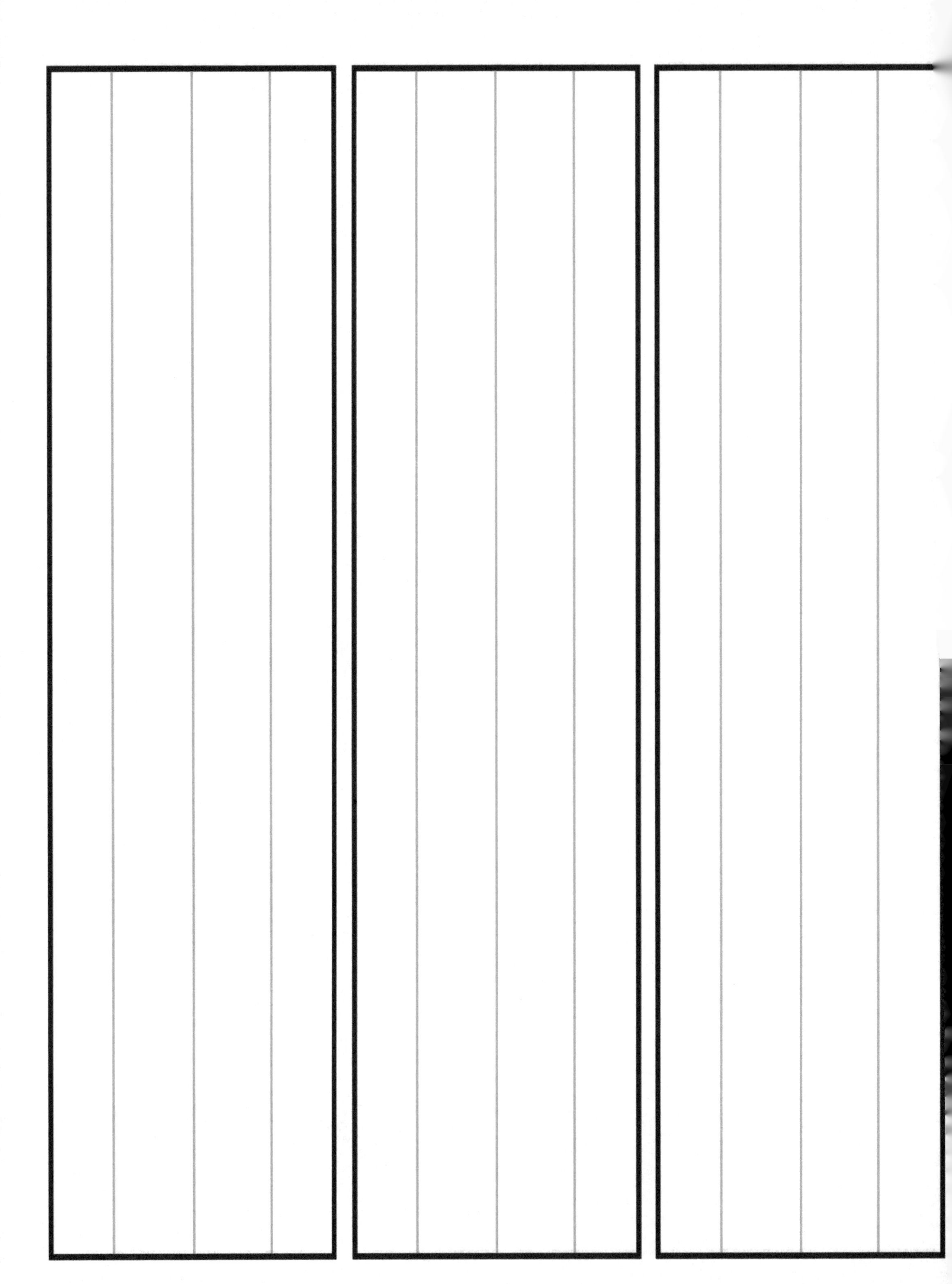

SUPER HERO ENGINEERS

Bridge Engineers really are super heroes 'cause they fight the EVIL forces of compression and tension when they design bridges. Whoa!!! That's crazy!!! WAIT!!! What IS "compression" and "tension" anyway?

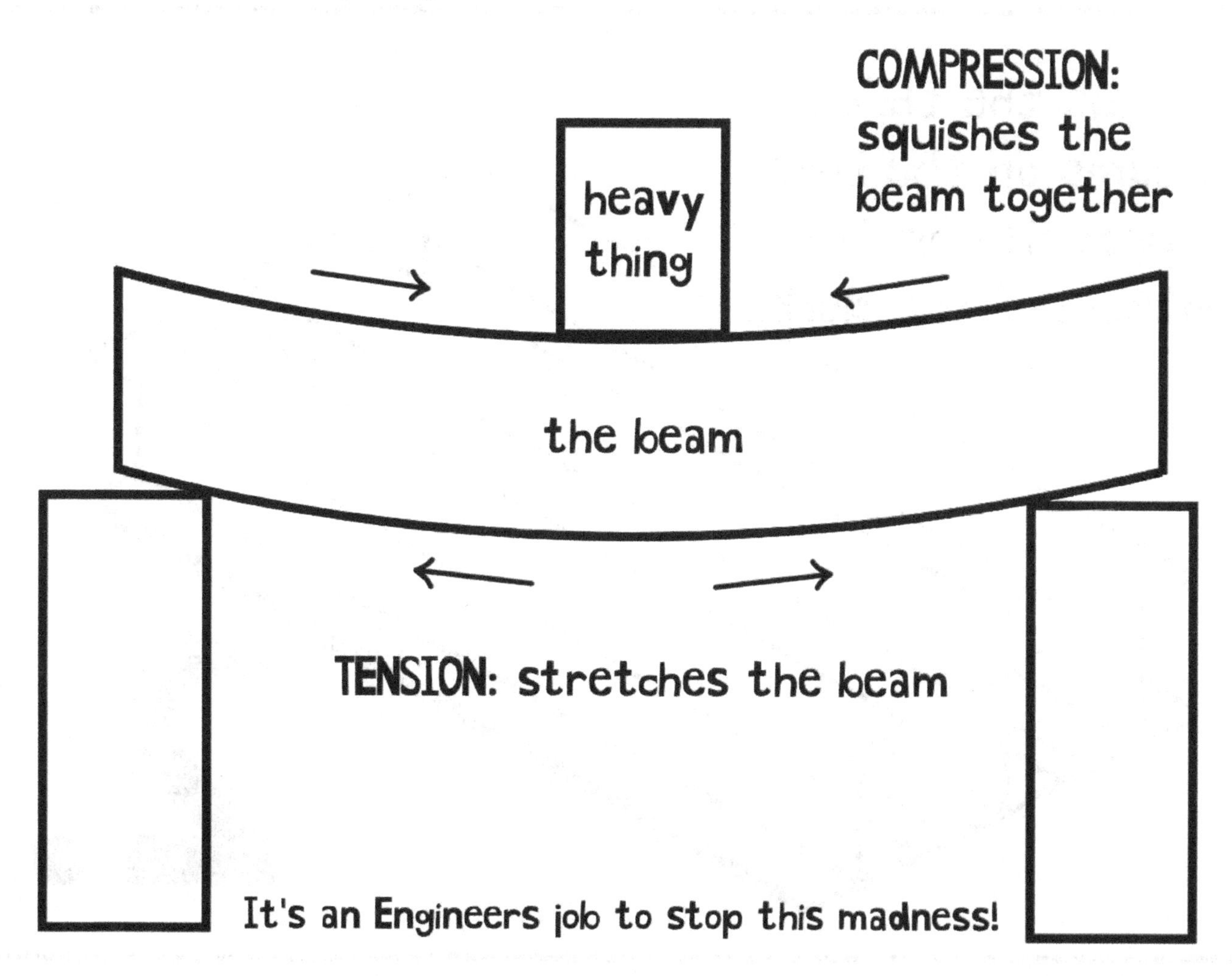

Cut out, fold, and glue the deck flat.

Glue the three beams on the lines provided on the bottom of the deck.

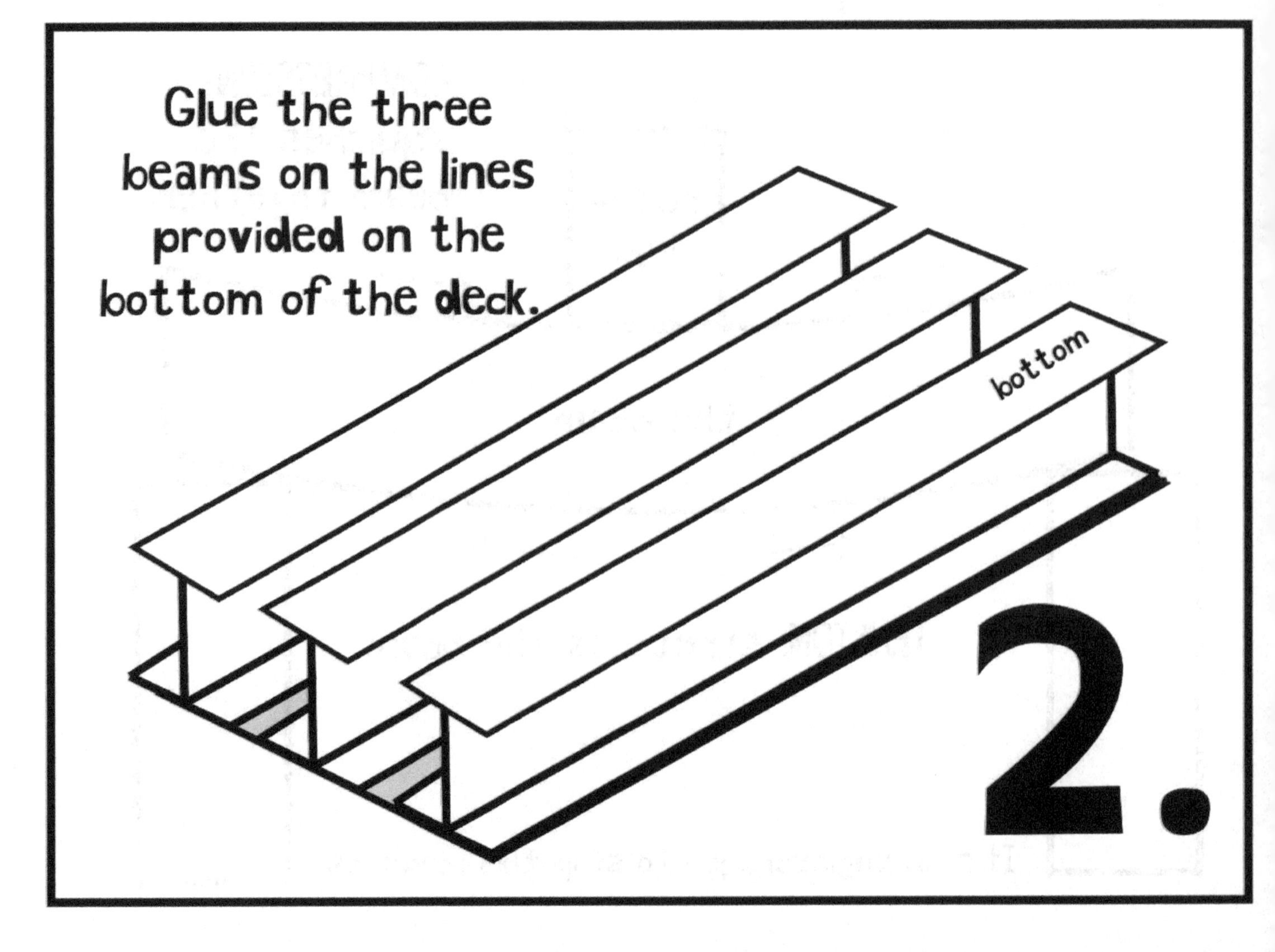

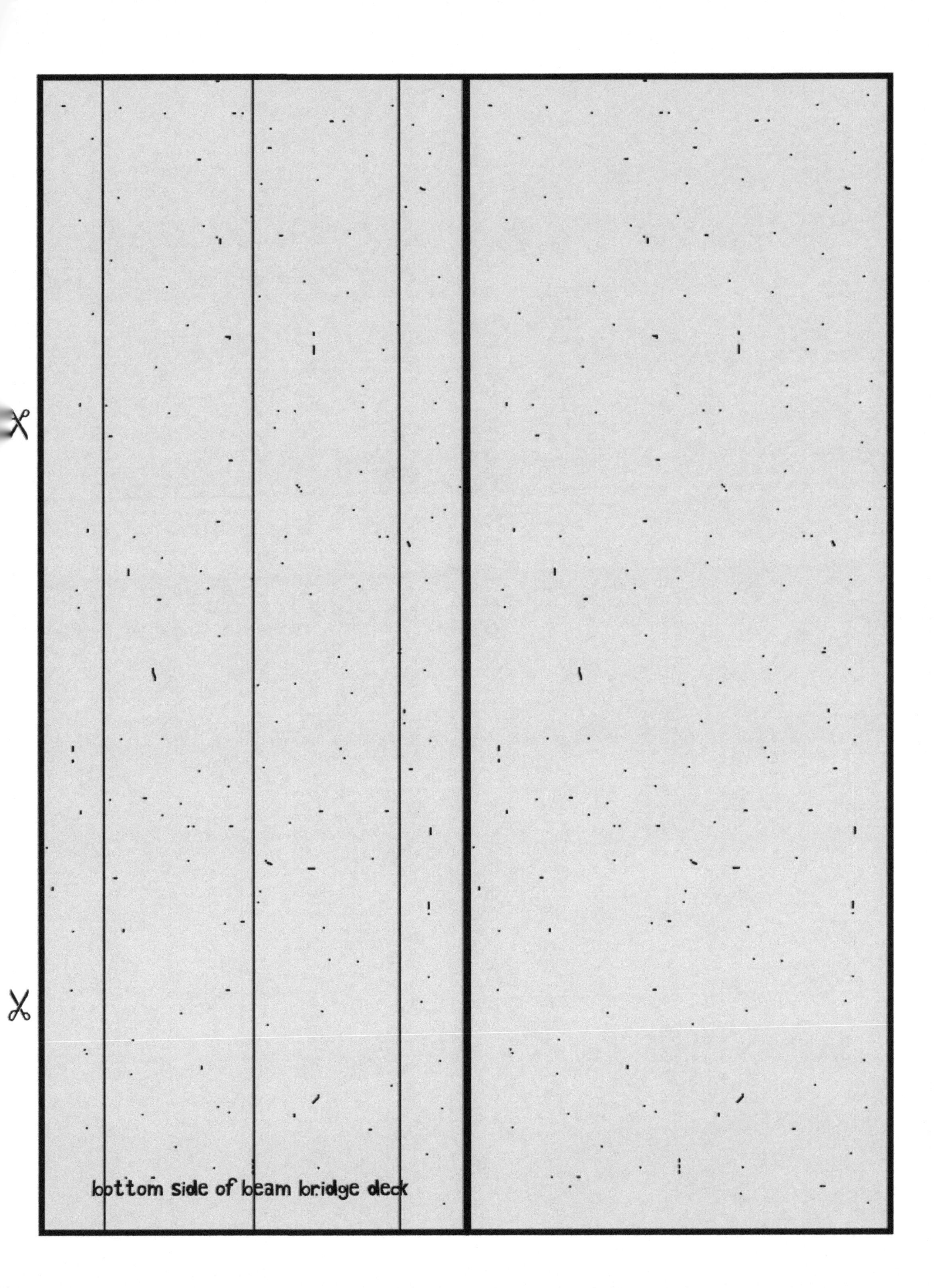
bottom side of beam bridge deck

bridge deck

fold on this line and glue

LET'S TEST OUR BRIDGE!

Engineers are always testing things to find out how they can improve their designs. Let's test our paper bridge. We're gonna need something like two stacks of books of equal height to put our bridge on. Then we need some weight. Coins work nicely! Record the results.

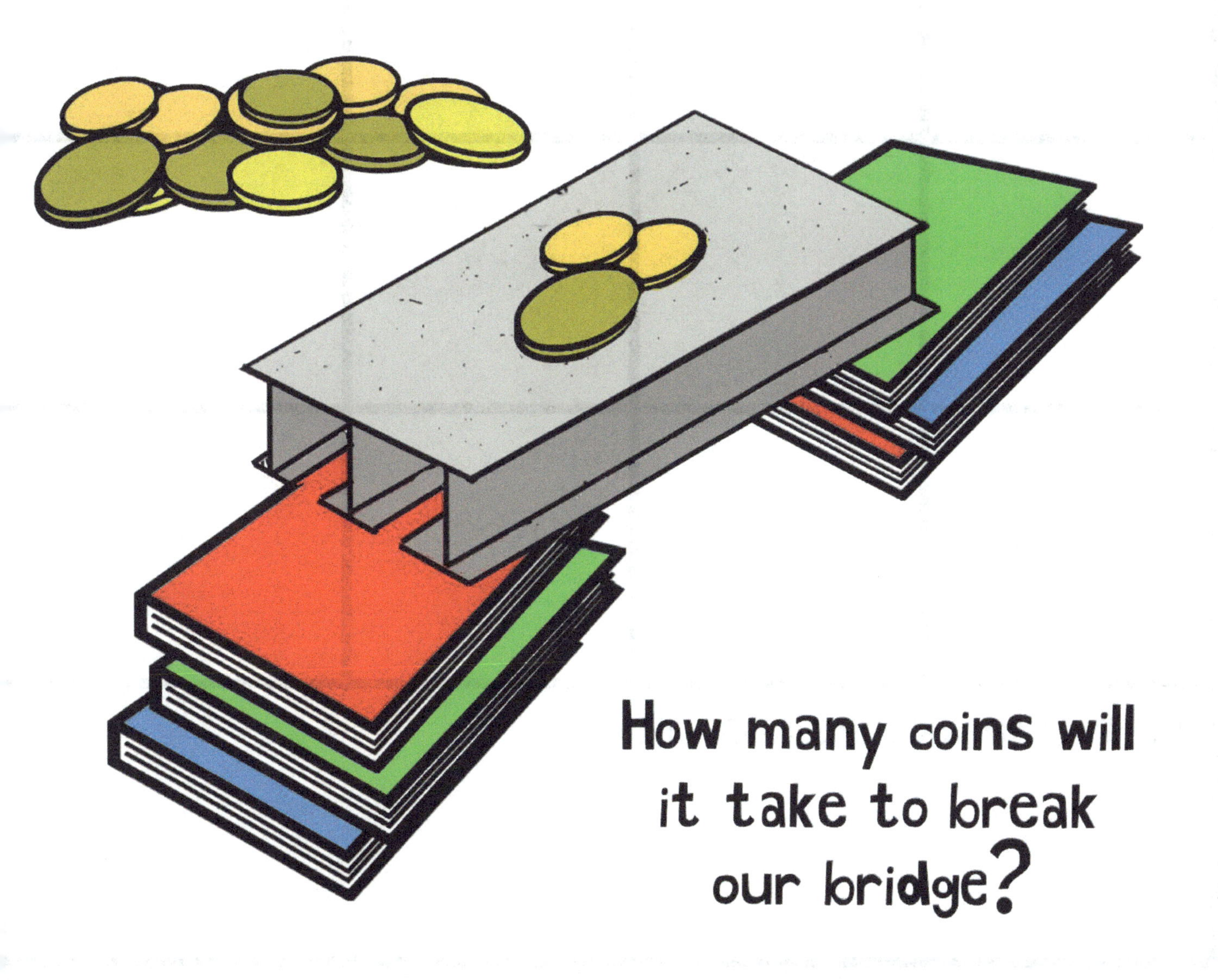

How many coins will it take to break our bridge?

BRIDGE TEST RESULTS

NUMBER OF COINS	DID IT SAG?	DID IT TWIST?	DID IT BREAK?

HOW WOULD YOU MAKE THIS BRIDGE STRONGER?

AEROSPACE ENGINEERING

Designing & Testing Airplane Wings

EXPERIMENT 1
AIR PRESSURE ON WINGS

What we need!

2 sheets of paper

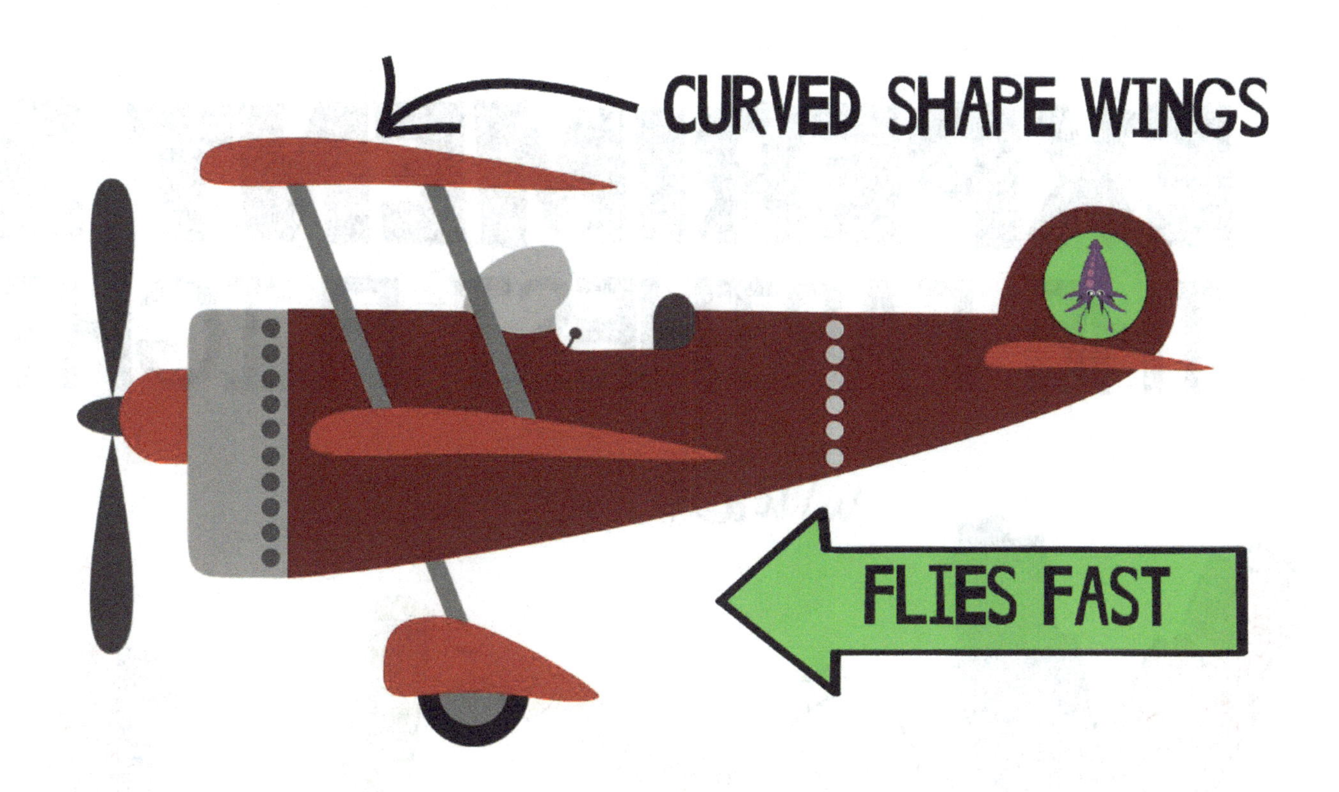

The high air pressure pushes up and the low air pressure pulls up the airplane.

THE EXPERIMENT

Hold the two pieces of paper parallel about 6 inches (15 cm) apart. Blow through the middle. What happened? High pressure is on the outside, low pressure on the inside making the papers move inward. Similar to a plane's wings.

EXPERIMENT 2
THE FARTHEST FLIGHT

PLANE 1

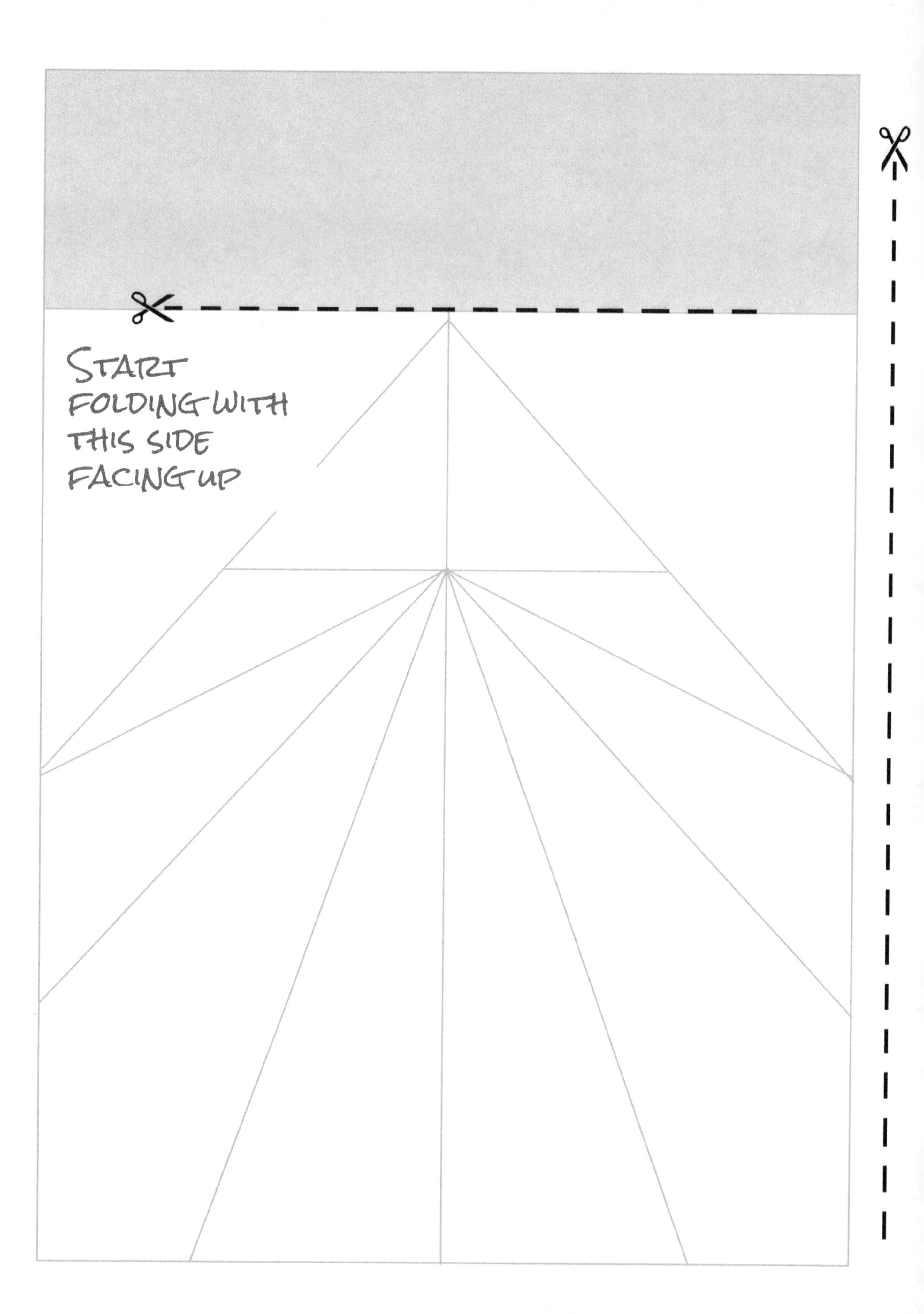
START
FOLDING WITH
THIS SIDE
FACING UP

BEFORE CUTTING OUT THE PLANES GIVE THEM SOME COLOR AND DESIGN. THE KEYS BELOW SHOULD HELP TO KNOW WHAT PART OF THE PLANE TO DRAW AND COLOR.

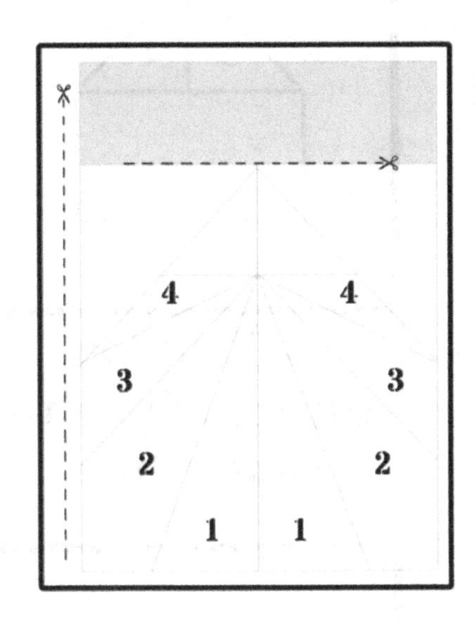

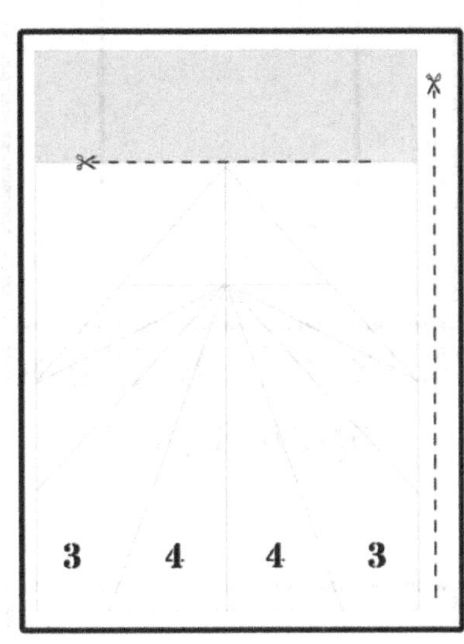

PLANE 1

1. Inner Side
2. Top Wing
3. Bottom Wing
4. Outer Side

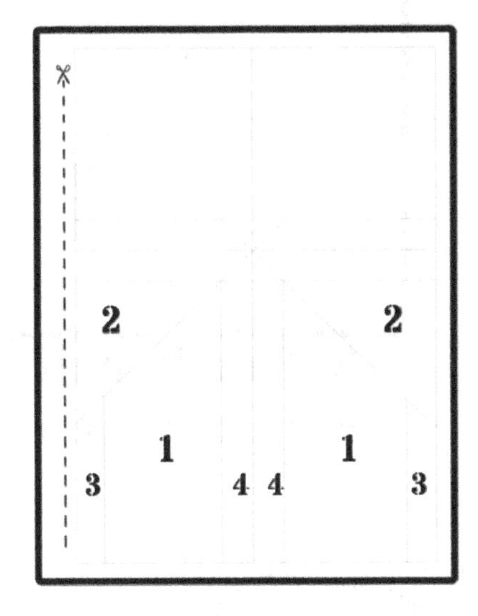

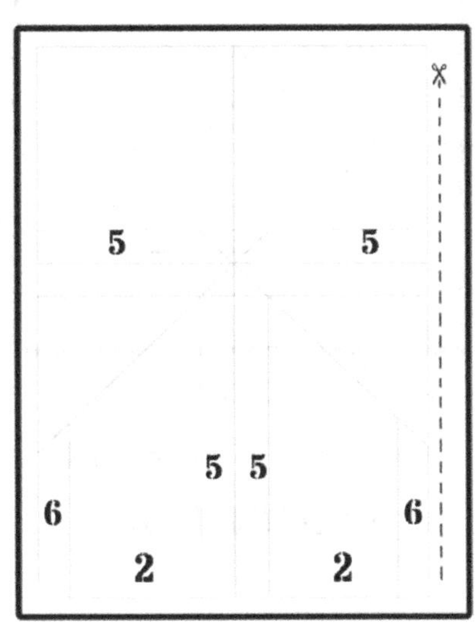

PLANE 2

1. Top Wing
2. Bottom Wing
3. Inner Wing Tip
4. Inner Side
5. Outer Side
6. Outer Wing Tip

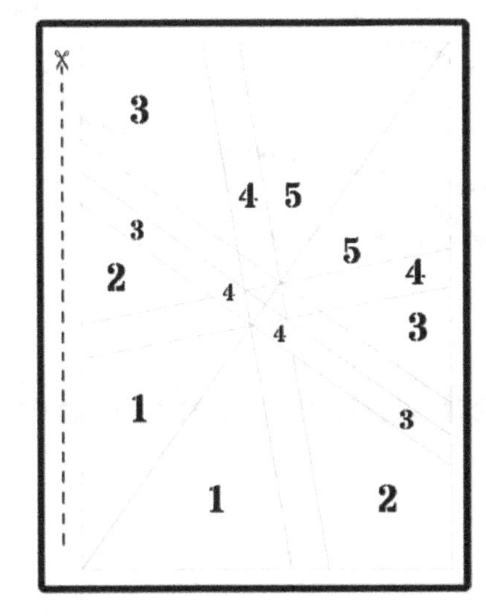

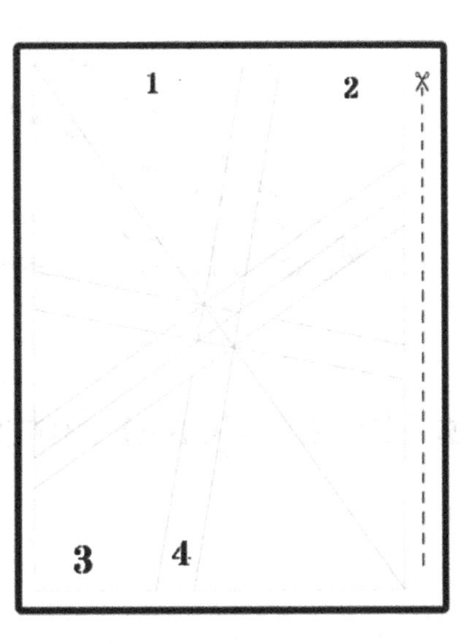

PLANE 3

1. Fin
2. Top Wing
3. Bottom Wing
4. Outer Side
5. Inner Side

PLANE 1

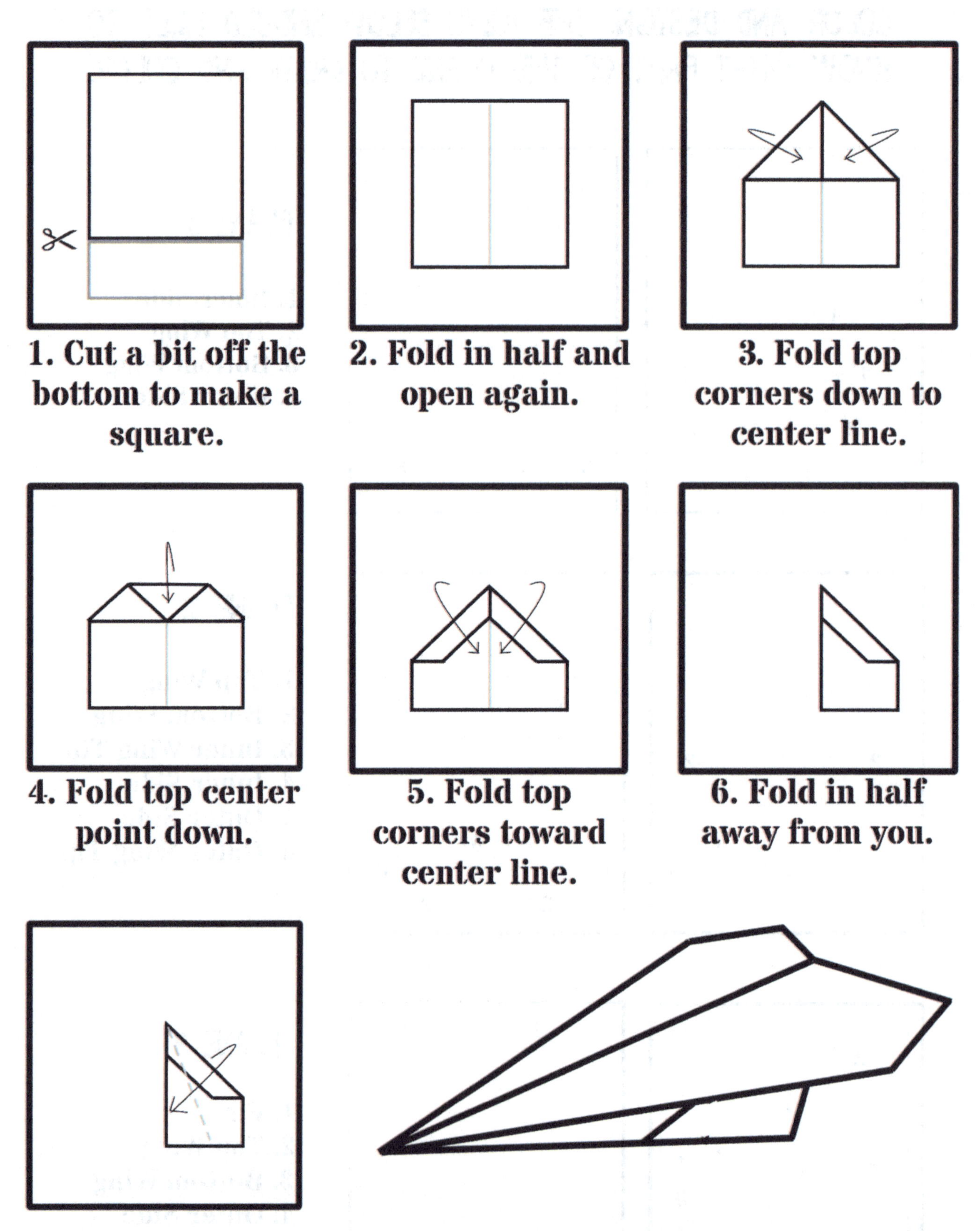

7. Fold wings down on the diagonal dotted line on both sides.

PLANE 2

START
FOLDING WITH
THIS SIDE
FACING UP

Plane 2

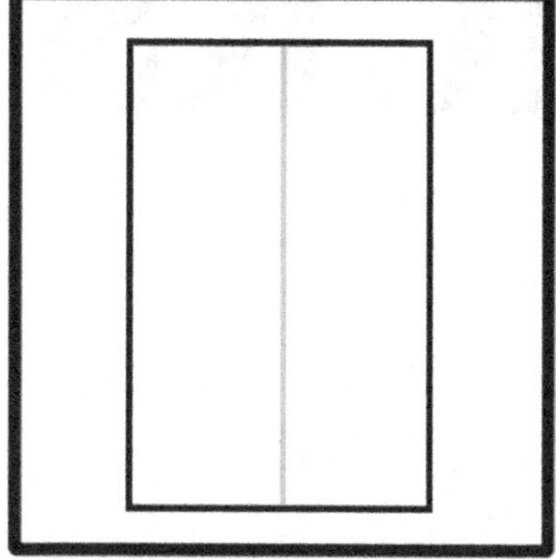
1. Fold paper in half and open again.

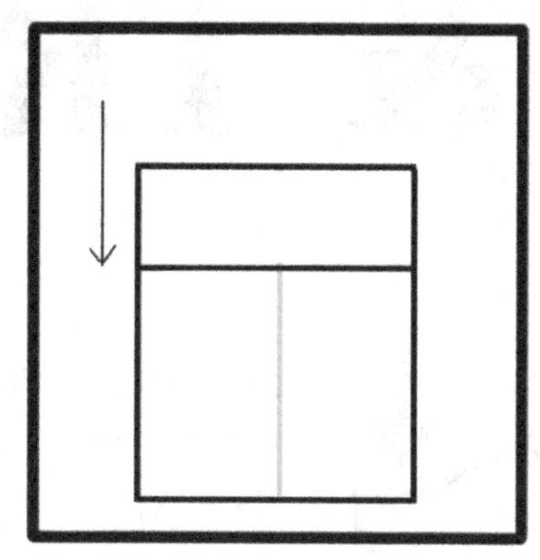
2. Fold top down.

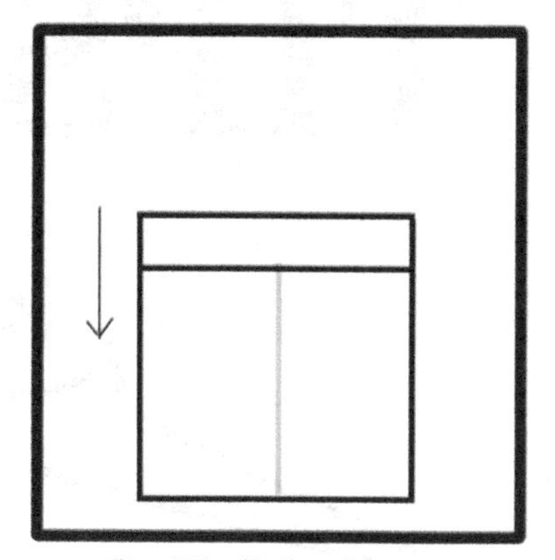
3. Fold top section in half.

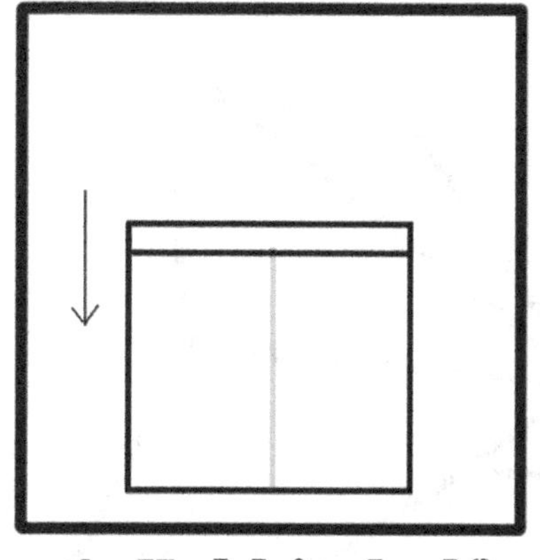
4. Fold in half again.

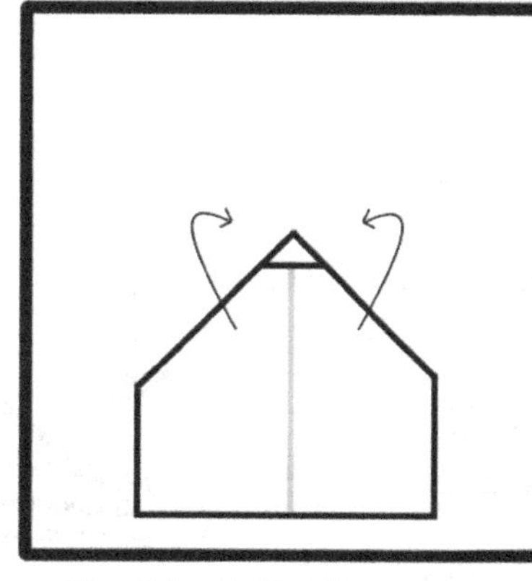
5. Fold the two corners back away from you.

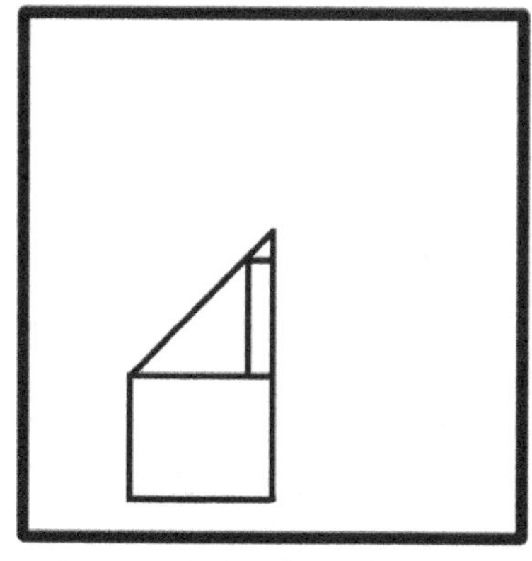
6. Fold in half.

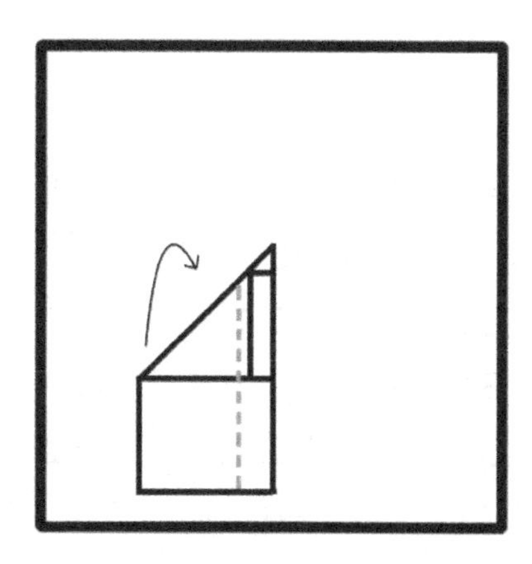
7. Fold wings down on dotted line.

8. Fold wing tips up.

TIPS & TRICKS

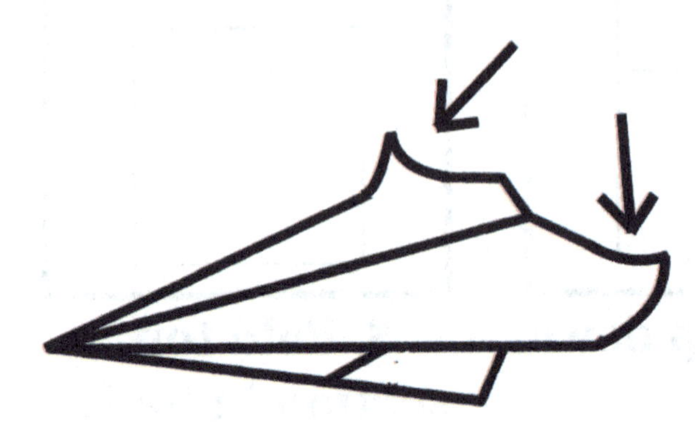

If you want the plane to go up, slightly bend the back wings up. If you want it to go down, bend them slightly down.

If you want the plane to go faster and further, add tape or a paper clip to keep it from opening and to add weight.

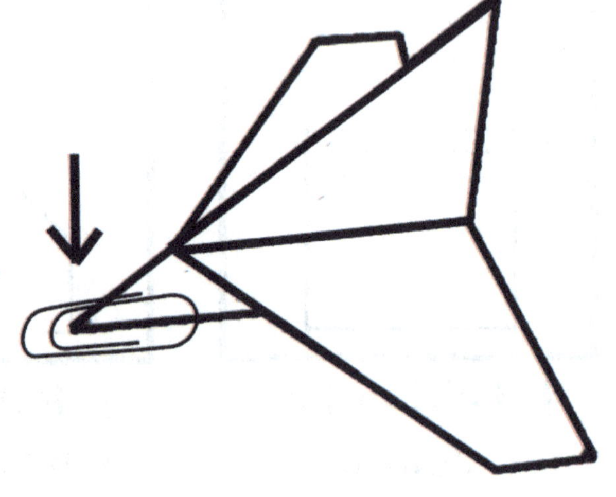

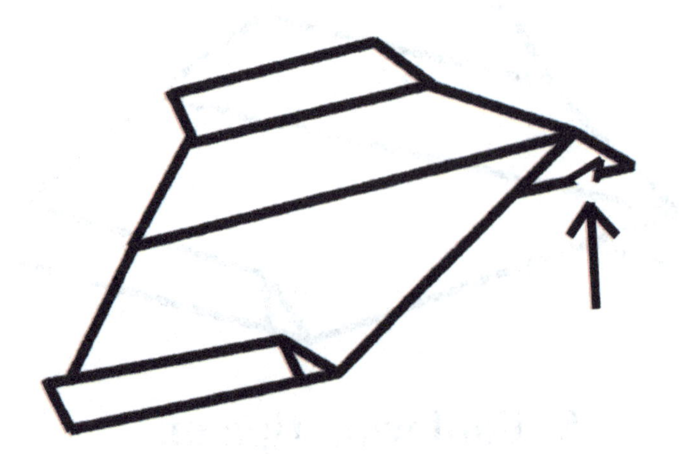

If you want a better launch, cut a small cut at the bottom of the nose and use a rubber band and your thumb to launch it.

PLANE 3

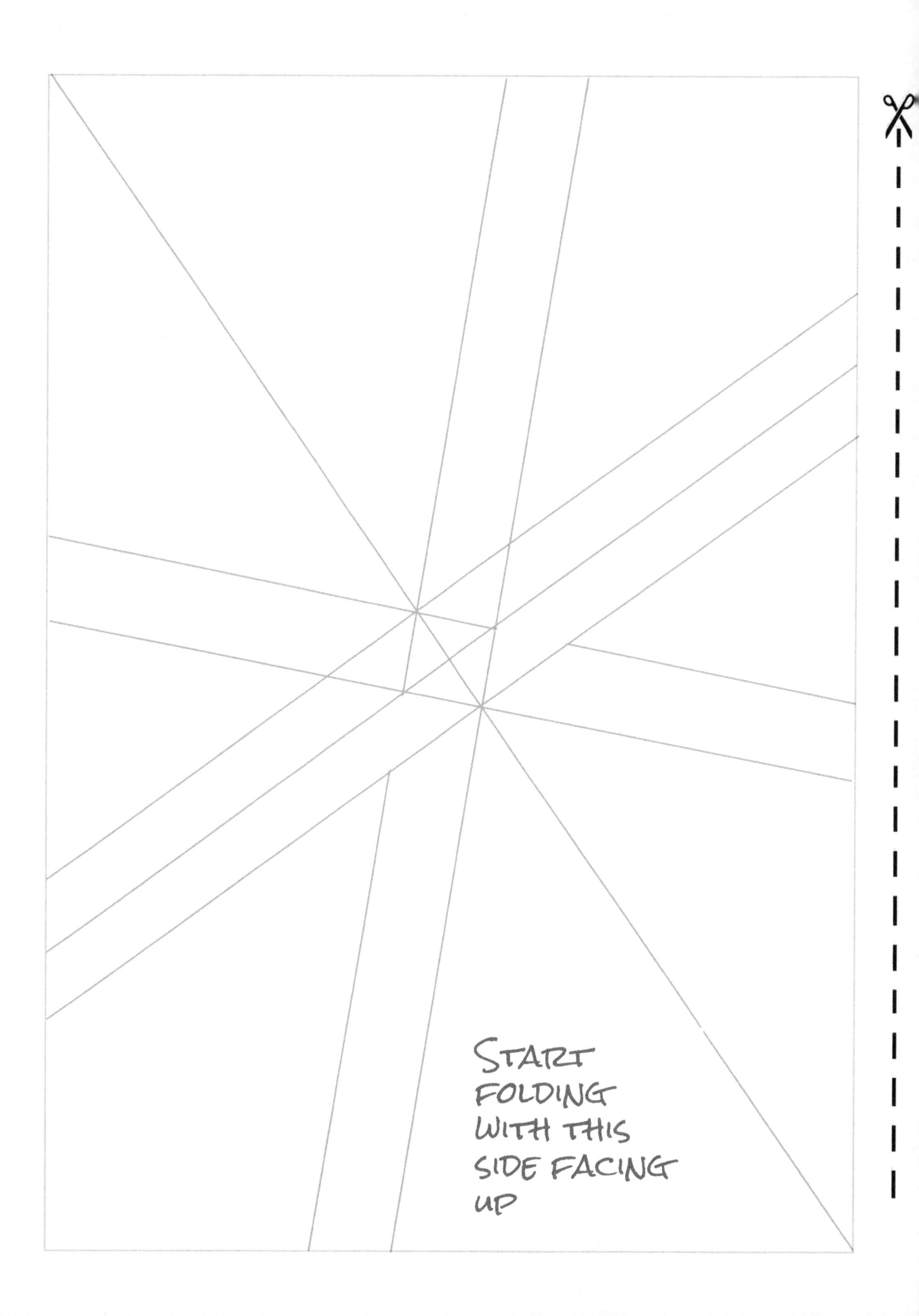
START FOLDING WITH THIS SIDE FACING UP

Plane 3

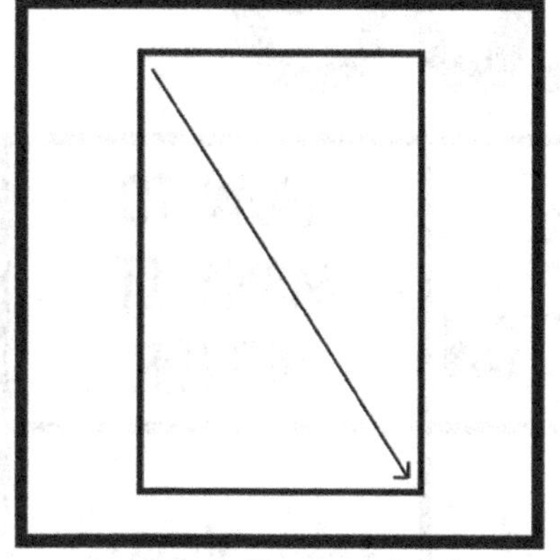

1. Fold top left corner to bottom right corner.

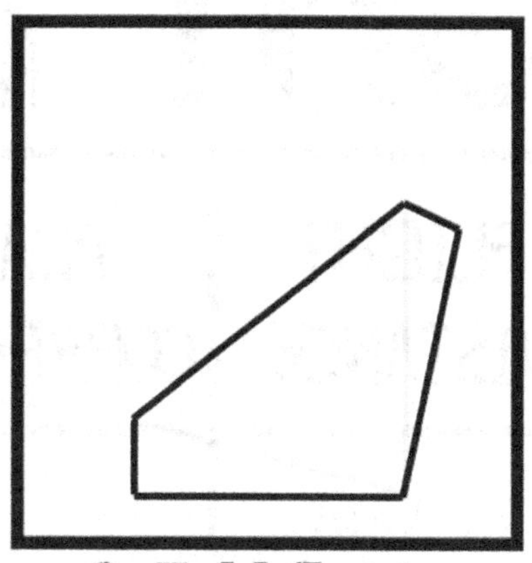

2. Fold flat to make this shape.

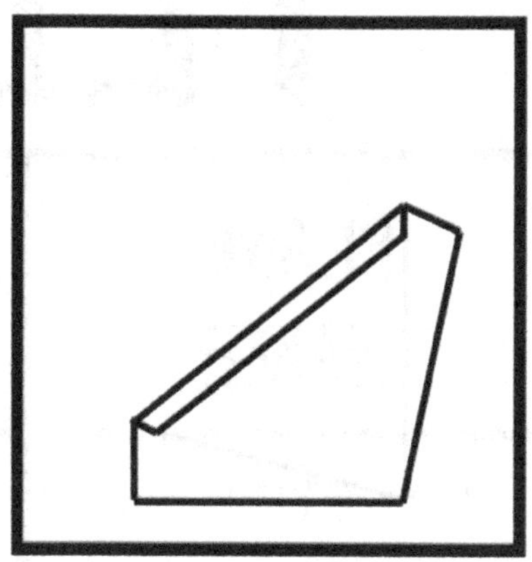

3. Fold over a small flap like this.

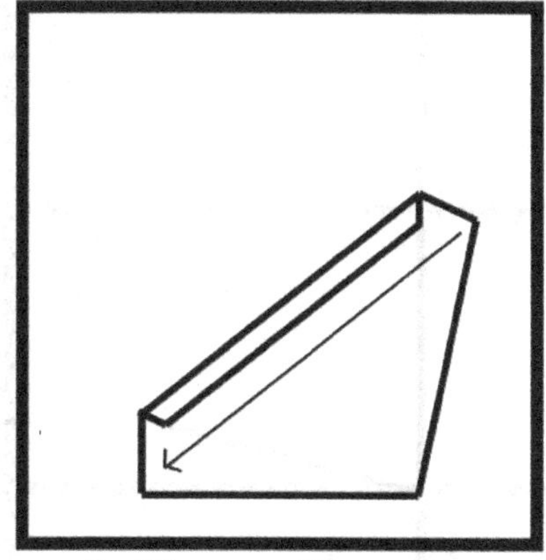

4. From the top right side, fold in half to the bottom left side.

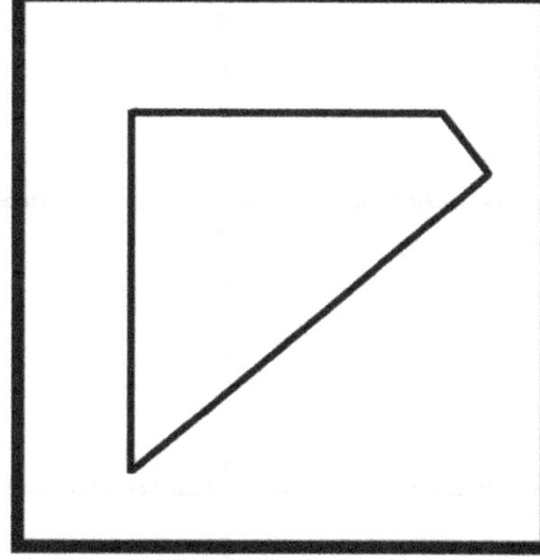

5. Flip over so the right angle is on the left.

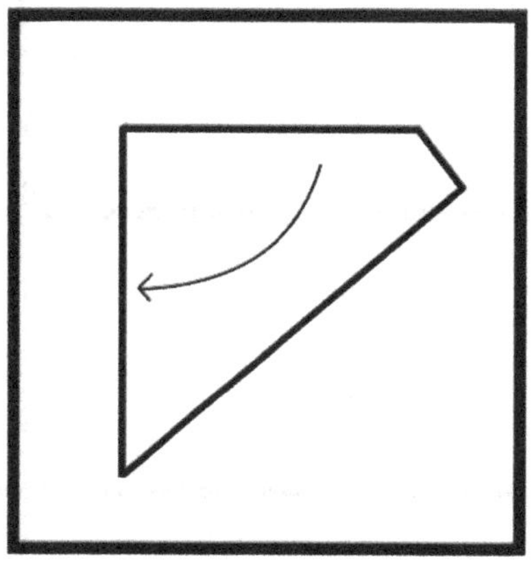

6. From the top right, fold one flap down.

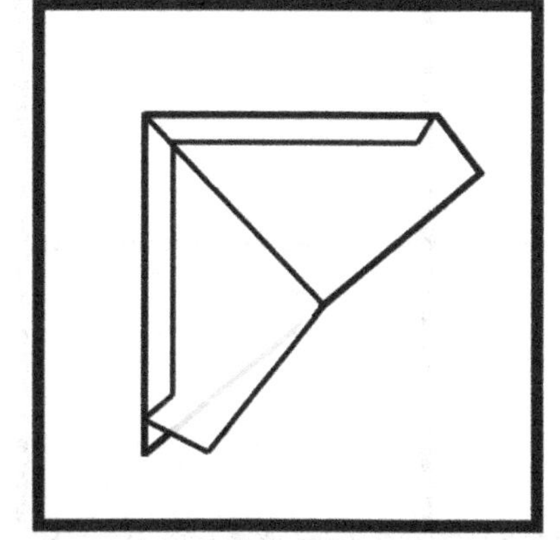

7. Fold flat to make this shape.

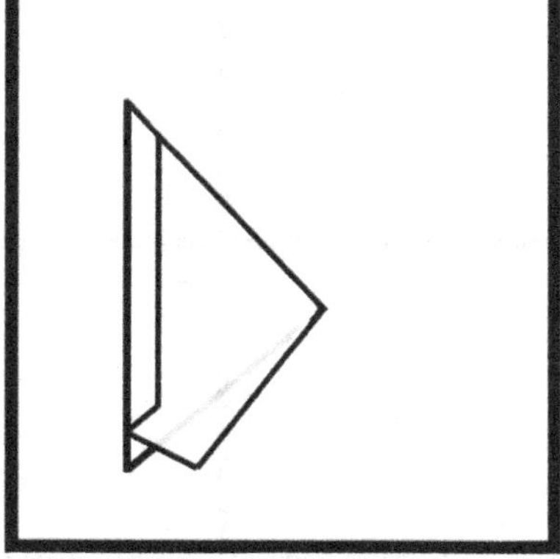

8. Fold the second flap down on the other side.

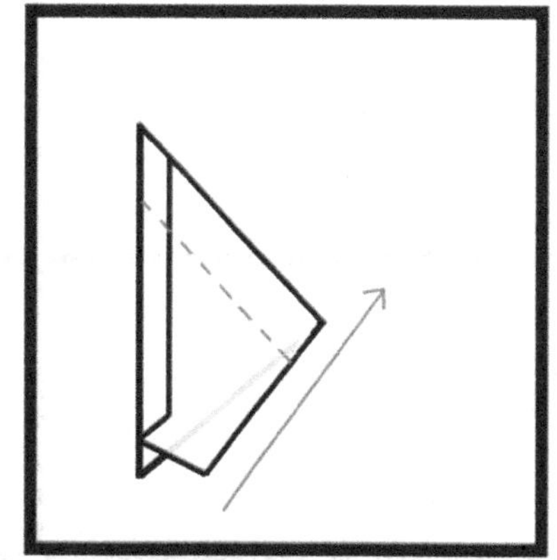

9. Fold wings on dotted line on both sides.

FLIGHT LOG

PLANE NAME	FLIGHT DISTANCE	PLANE CORRECTION	HOW TO MAKE IT BETTER
(example) plane 1	10 feet (3 meters)	up turned back wings	add weight to front

MECHANICAL ENGINEERING

How Machine Cranks & Gears Work.

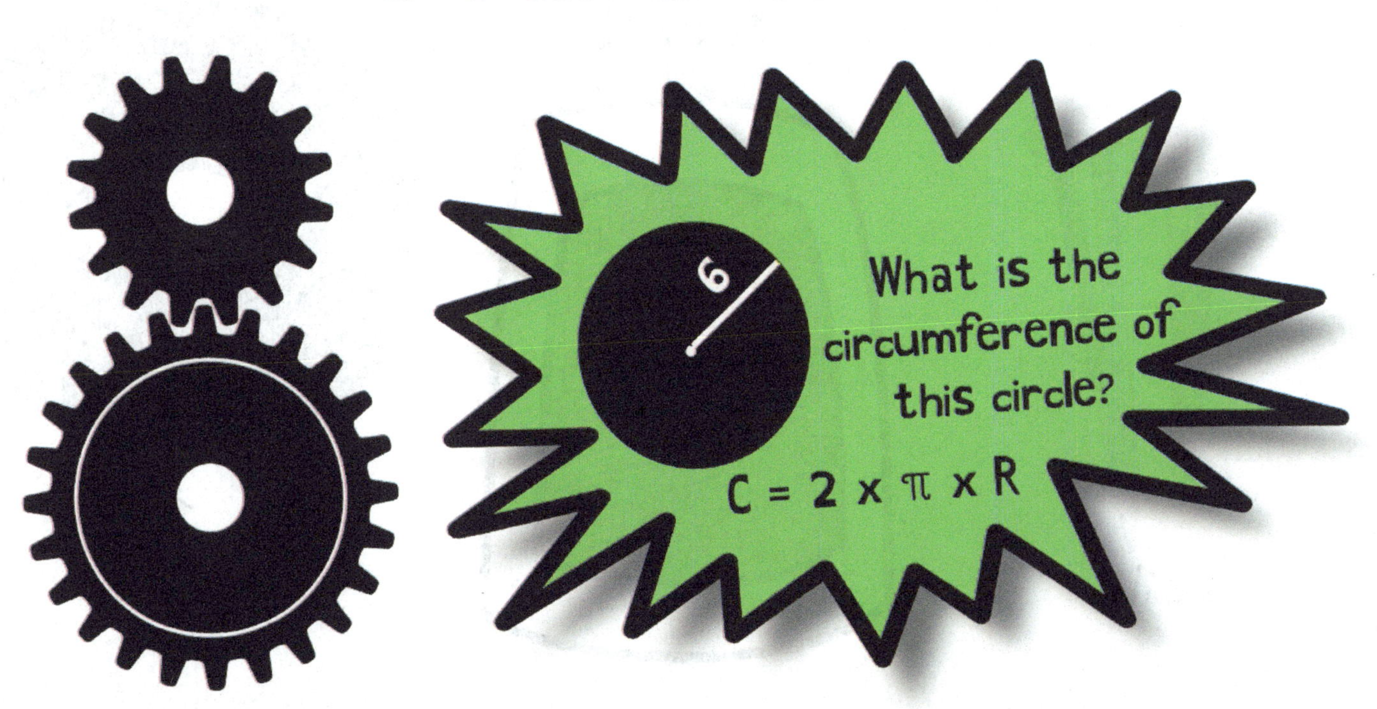

EXPERIMENT 1
BUILDING A GEAR TRAIN

What we need!

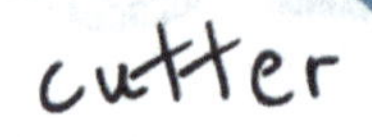

cardboard

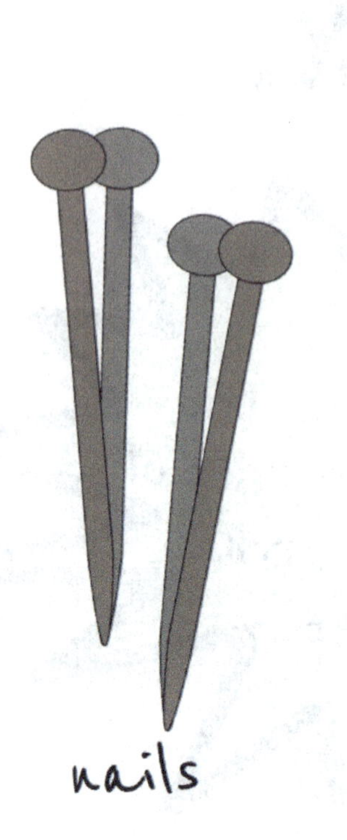

nails

tape

All About GEARS

When you put two or more gears together it's called a:

GEAR TRAIN

The gear with the force (crank) is called:

DRIVER

The gear that gets driven is the:

FOLLOWER

GEAR TRAIN WITH 3 OR MORE GEARS

Driver Idler Follower

When the **DRIVER** is smaller, less force is needed And the **FOLLOWER** spins slower.

Cut out the gear templates and glue them on to a cardboard sheet. Use the templates as a guide to cut around with the cutter.

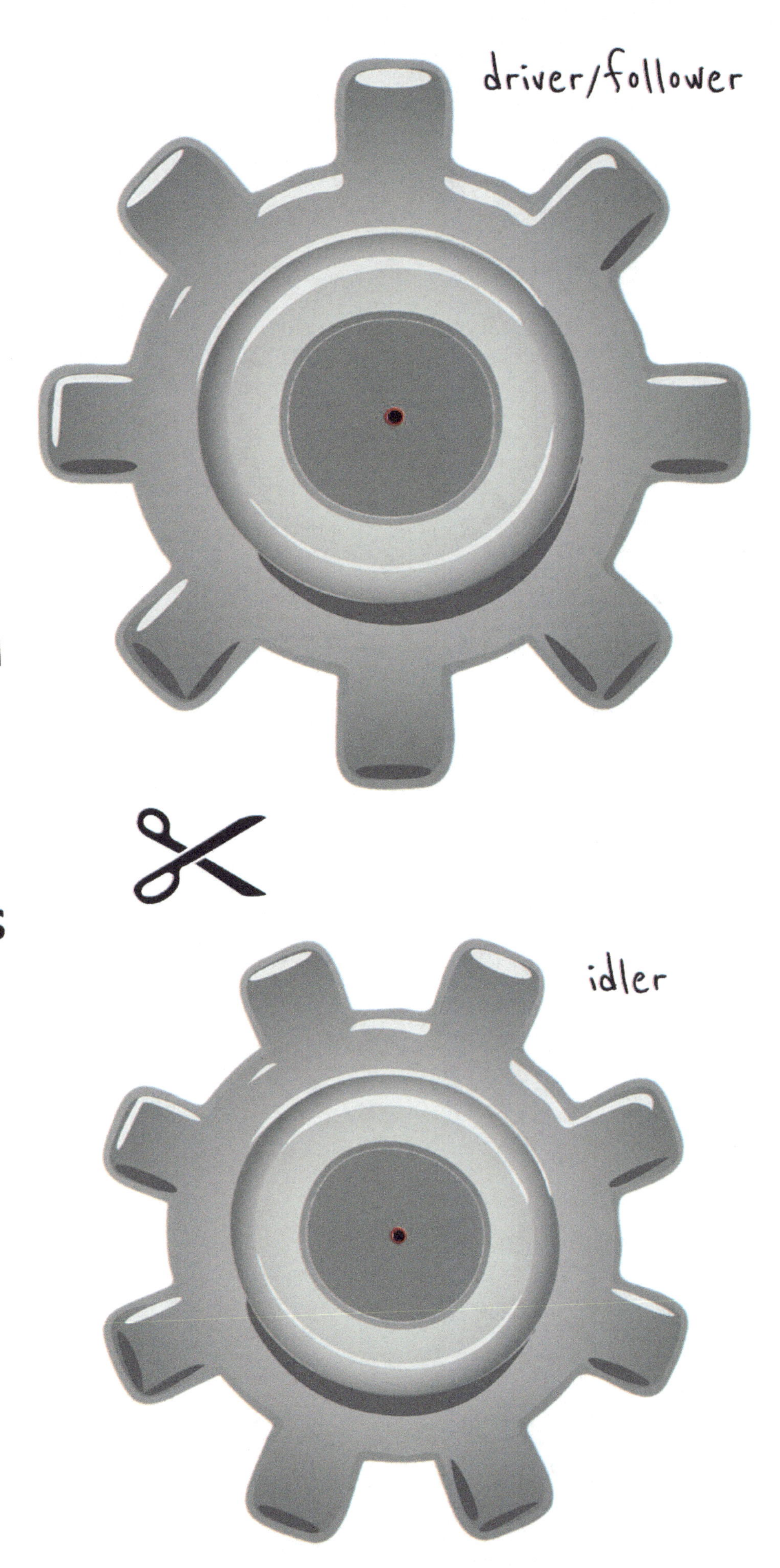

Cut out the gear templates and glue them on to a cardboard sheet. Use the templates as a guide to cut around with the cutter.

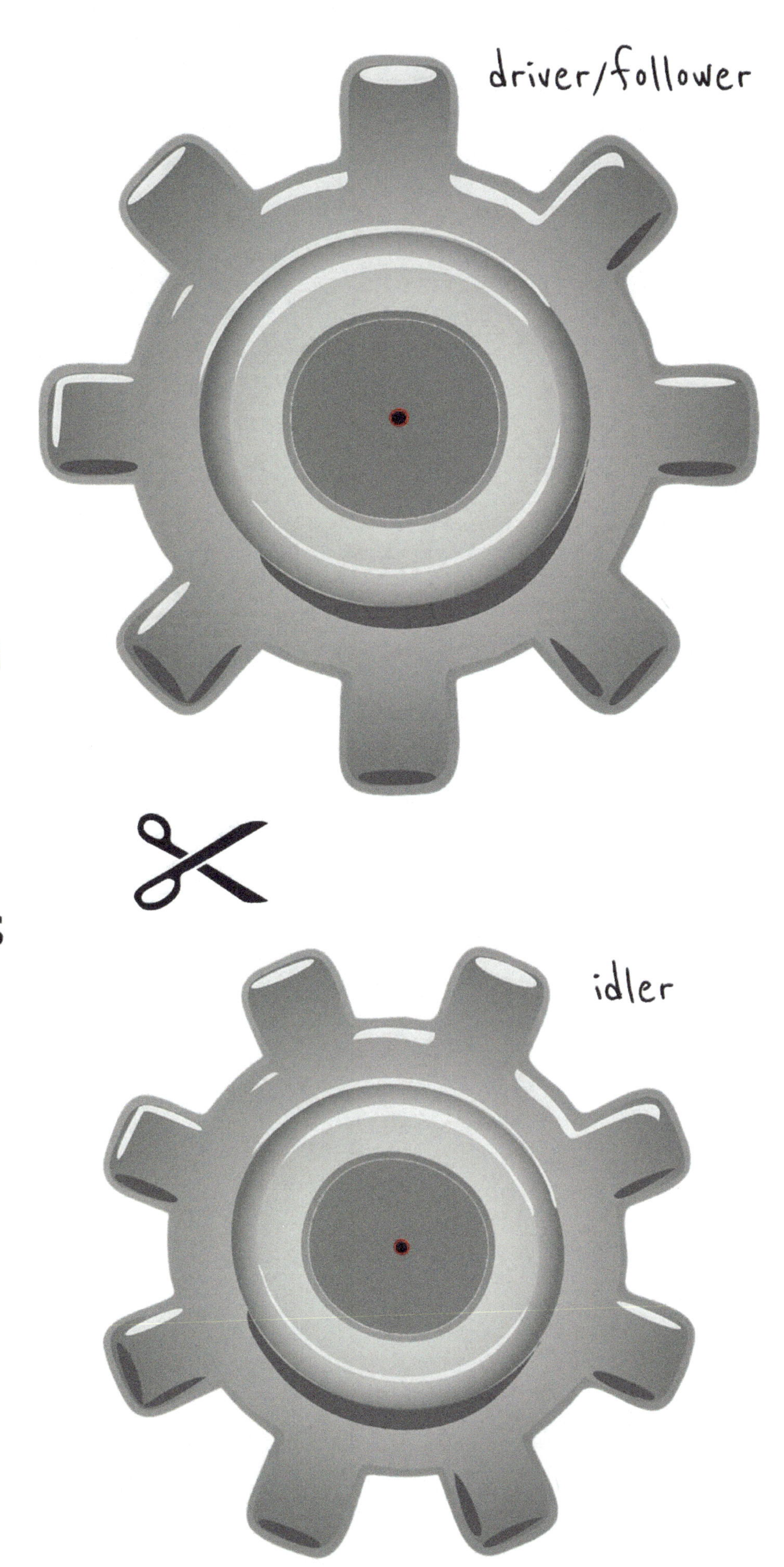

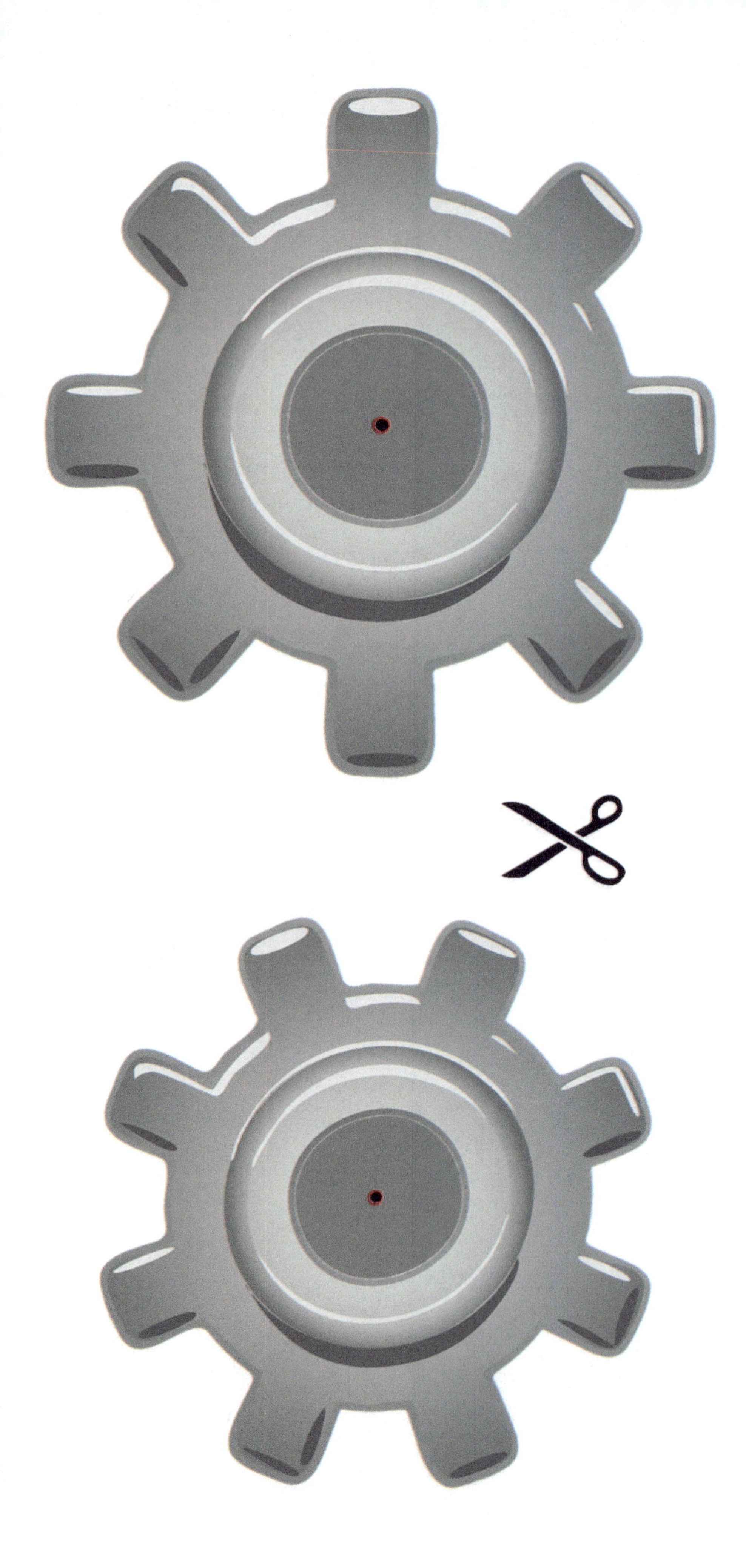

On another sheet of cardboard line up the gears. Once lined up, poke a hole through the middle of each gear with a nail and then through the sheet. Wrap tape around the nail to stop it from slipping out. Turn the driver gear. What happens?

GEAR DATA

WHAT DID YOU FIND OUT?

WHAT WOULD HAPPEN IF YOU ADDED ANOTHER GEAR?

WHAT WOULD YOU INVENT THAT USES GEARS?

EXPERIMENT 2
LET'S BUILD A GEARBOX

What we need!

cardboard

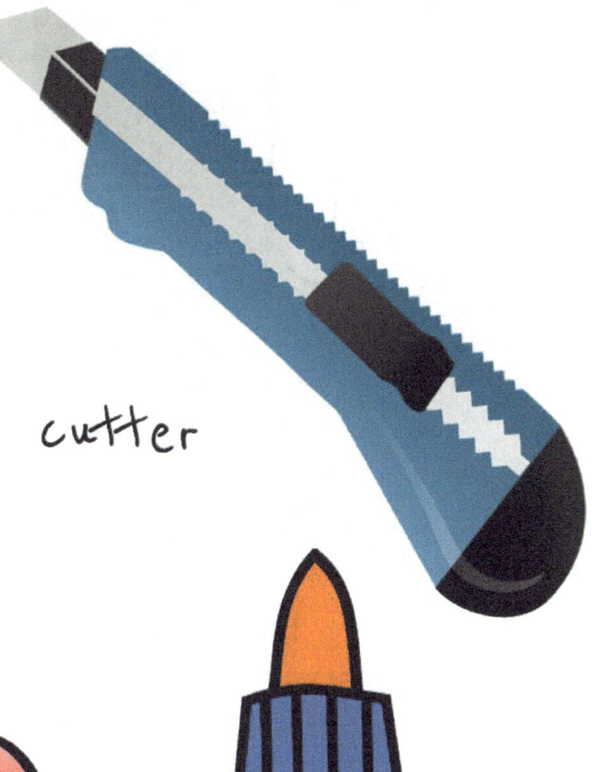

cutter

two long pencils
or pens

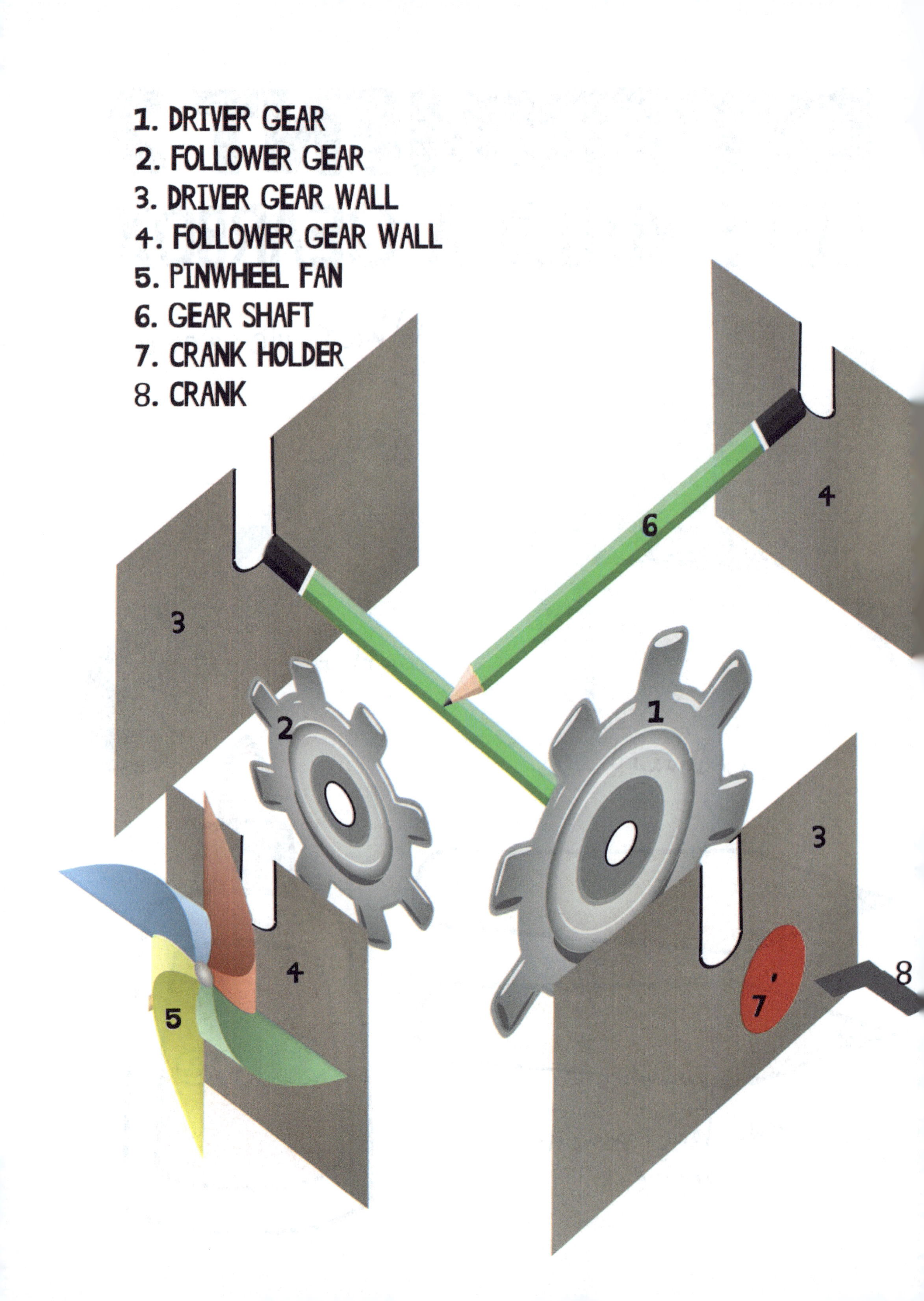
1. DRIVER GEAR
2. FOLLOWER GEAR
3. DRIVER GEAR WALL
4. FOLLOWER GEAR WALL
5. PINWHEEL FAN
6. GEAR SHAFT
7. CRANK HOLDER
8. CRANK
4
6
3
1
2
3
4
8
7
5

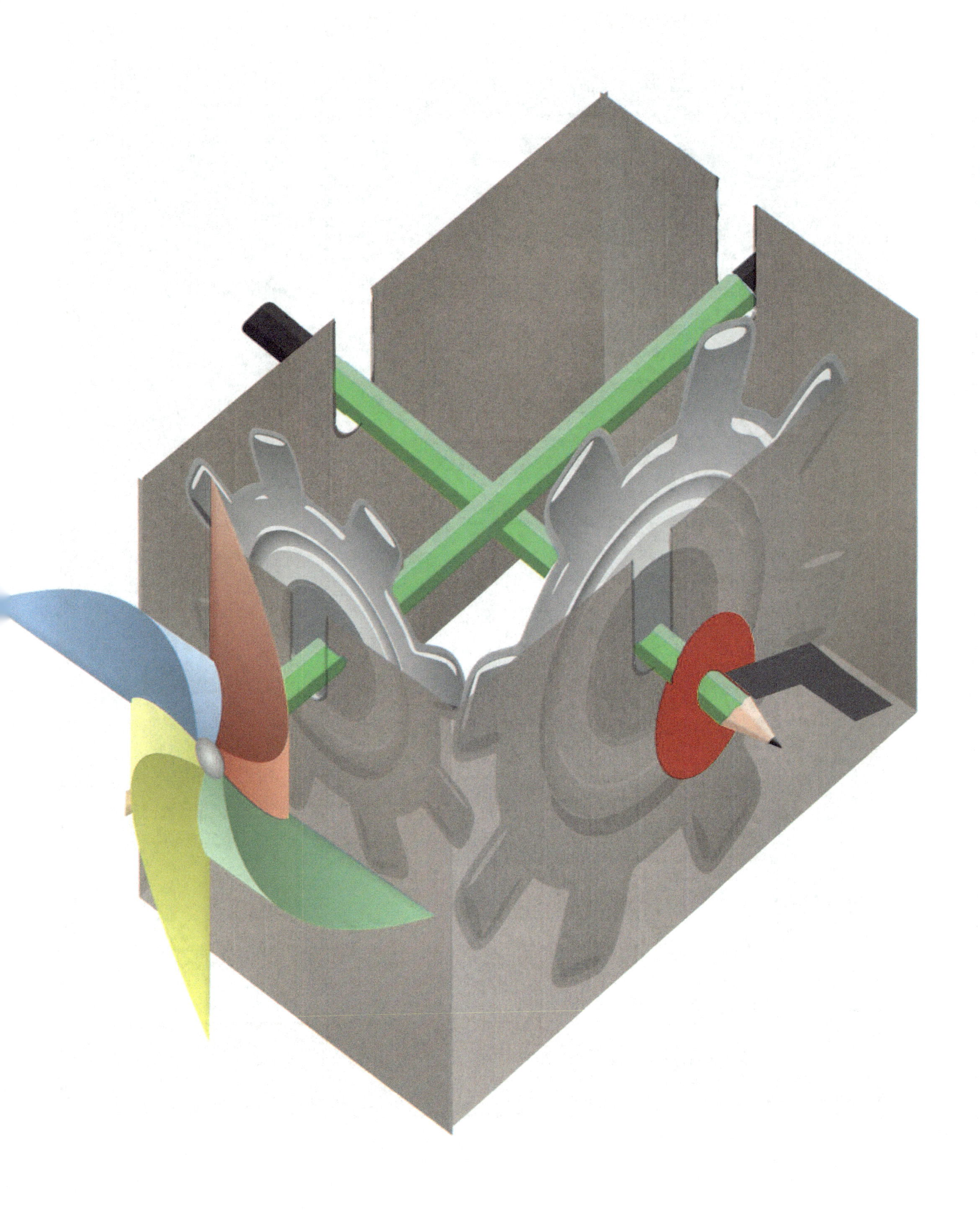

ASSEMBLY

1. Glue the 4 box walls.
2. Poke holes in the driver gear & crank holder.
3. With the second pencil poke a hole in the follower gear.
4. Place the driver gear & shaft in the wall slots.
5. Place the follower gear & shaft in wall slots and align gear teeth with the driver gear.
6. Once everything is in place, glue.
7. Glue the pinwheel.
8. Make adjustments as necessary.

These are the templates for the driver gear walls. Cut them out, glue on to a sheet of cardboard and then cut out the cardboard in these shapes.

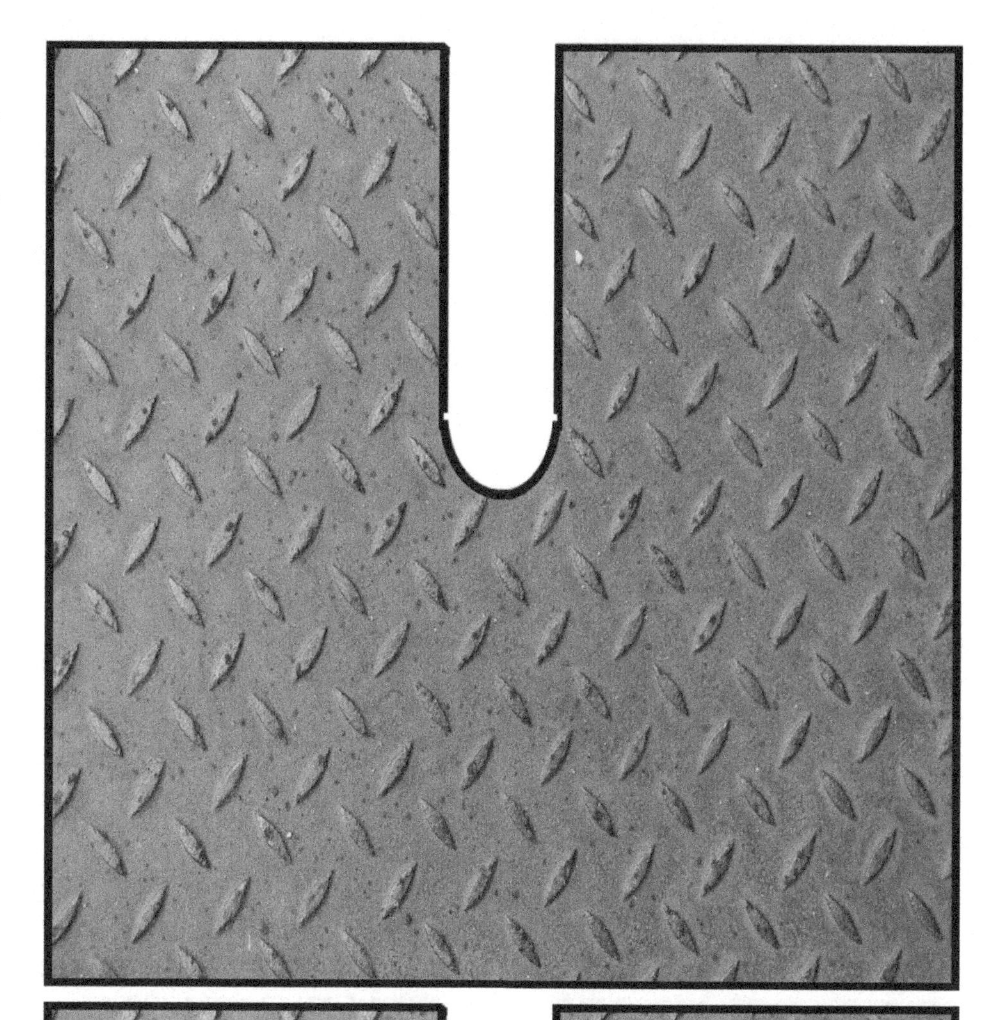

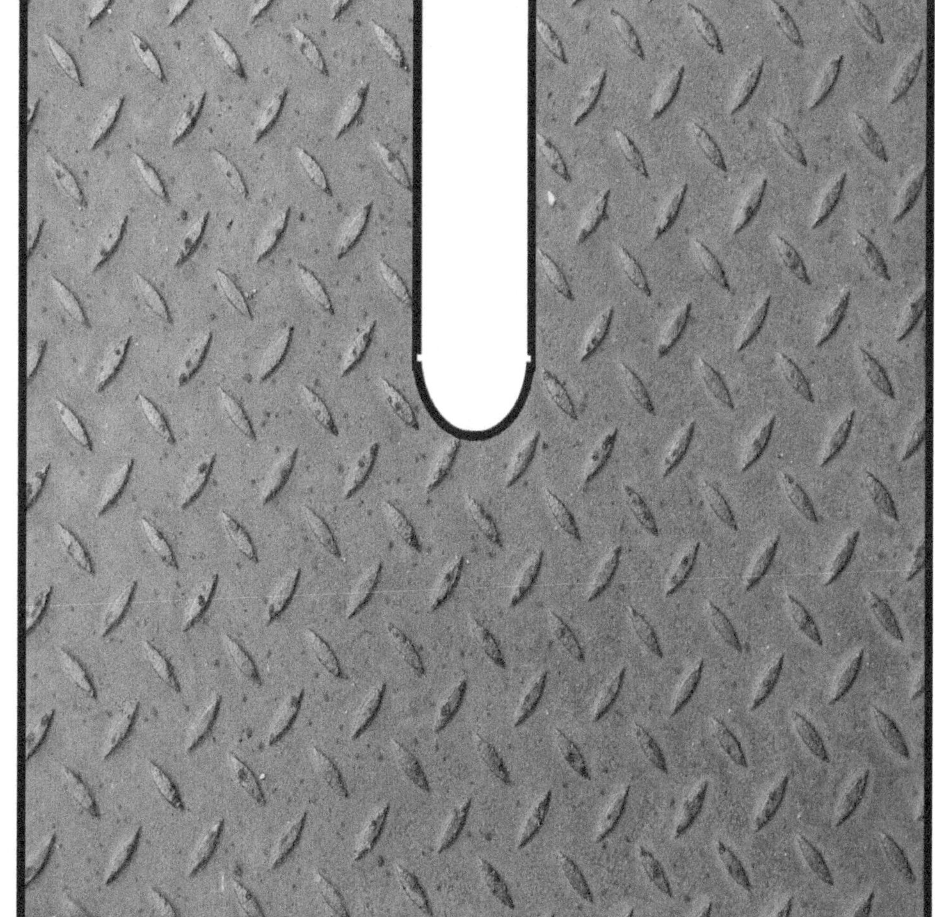

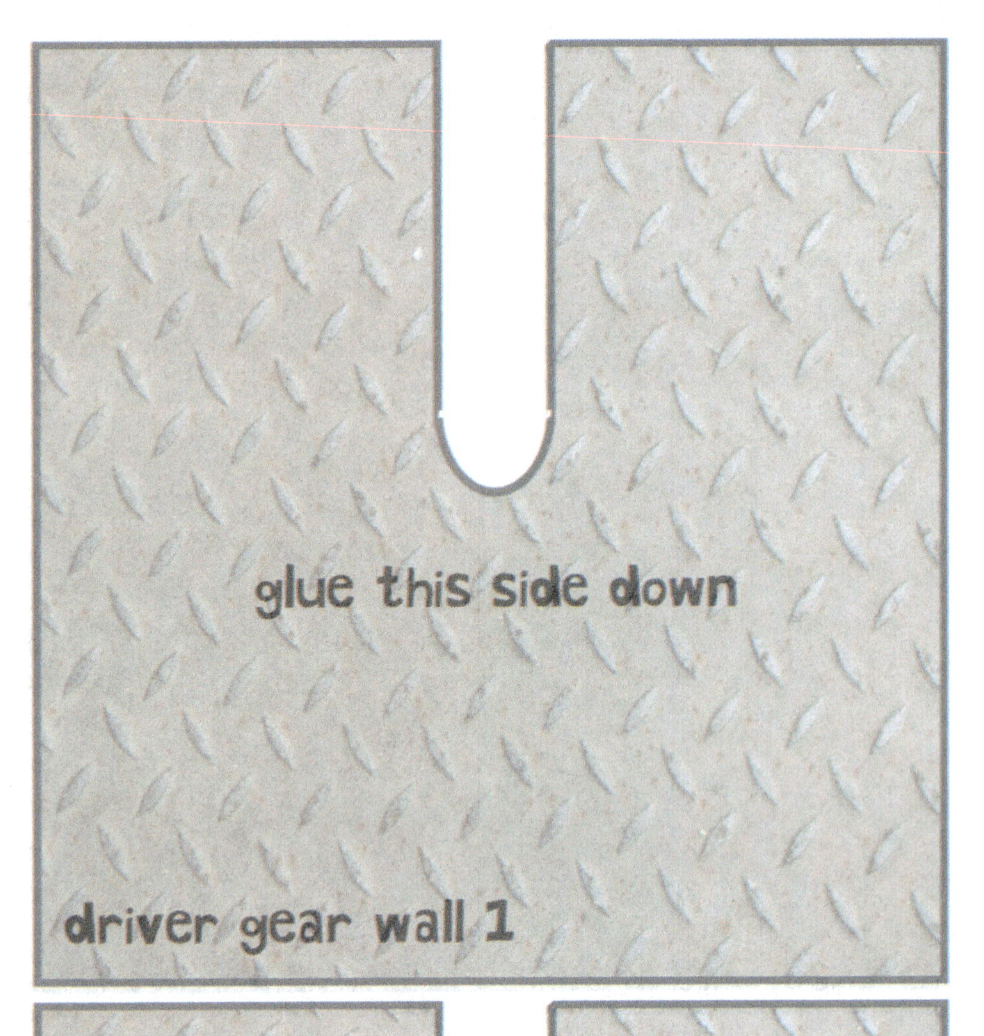

glue this side down

driver gear wall 2

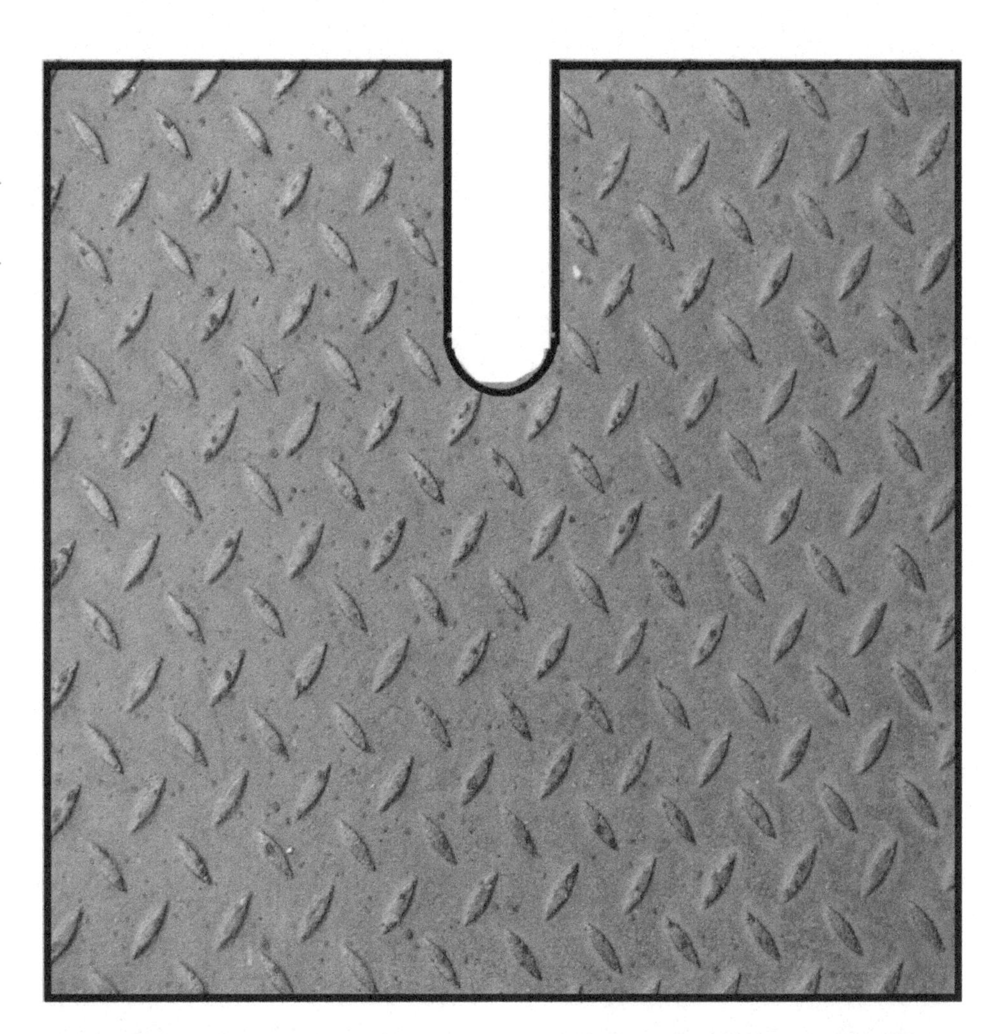

These are the templates for the follower gear walls. Cut them out, glue on to a sheet of cardboard and then cut out the cardboard in these shapes.

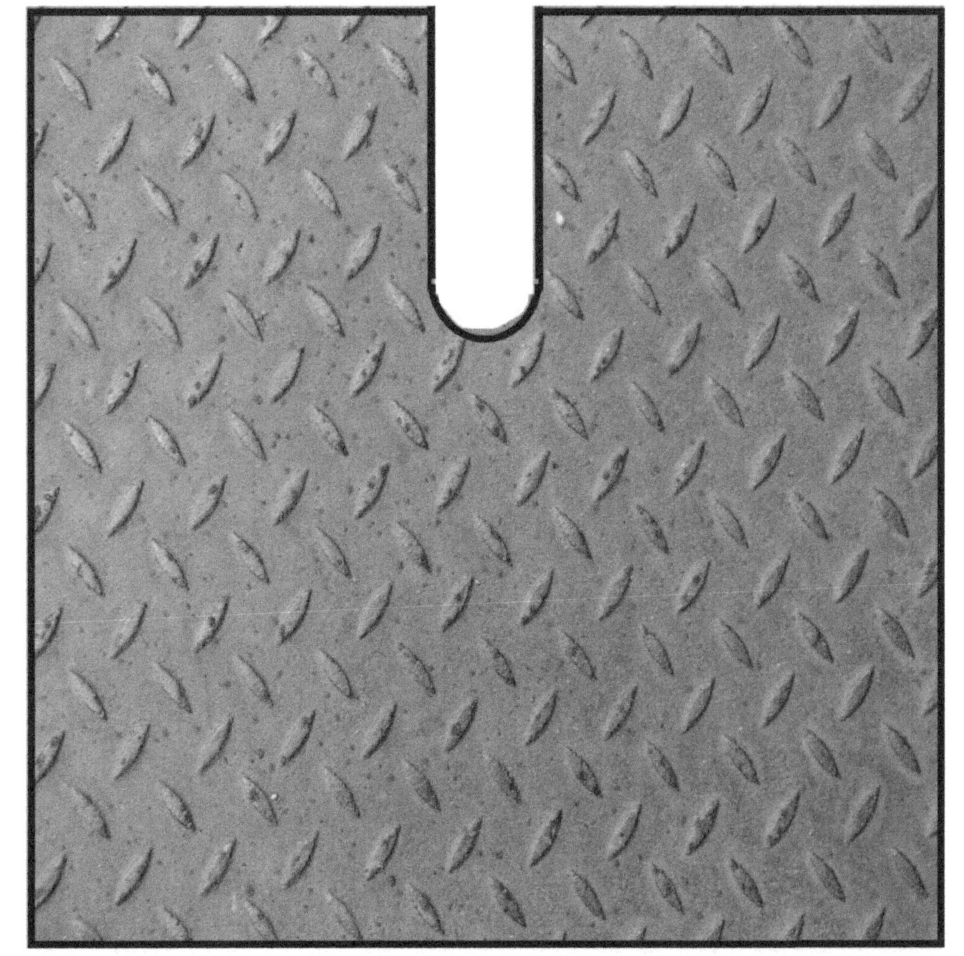

glue this side down
follower gear wall 1

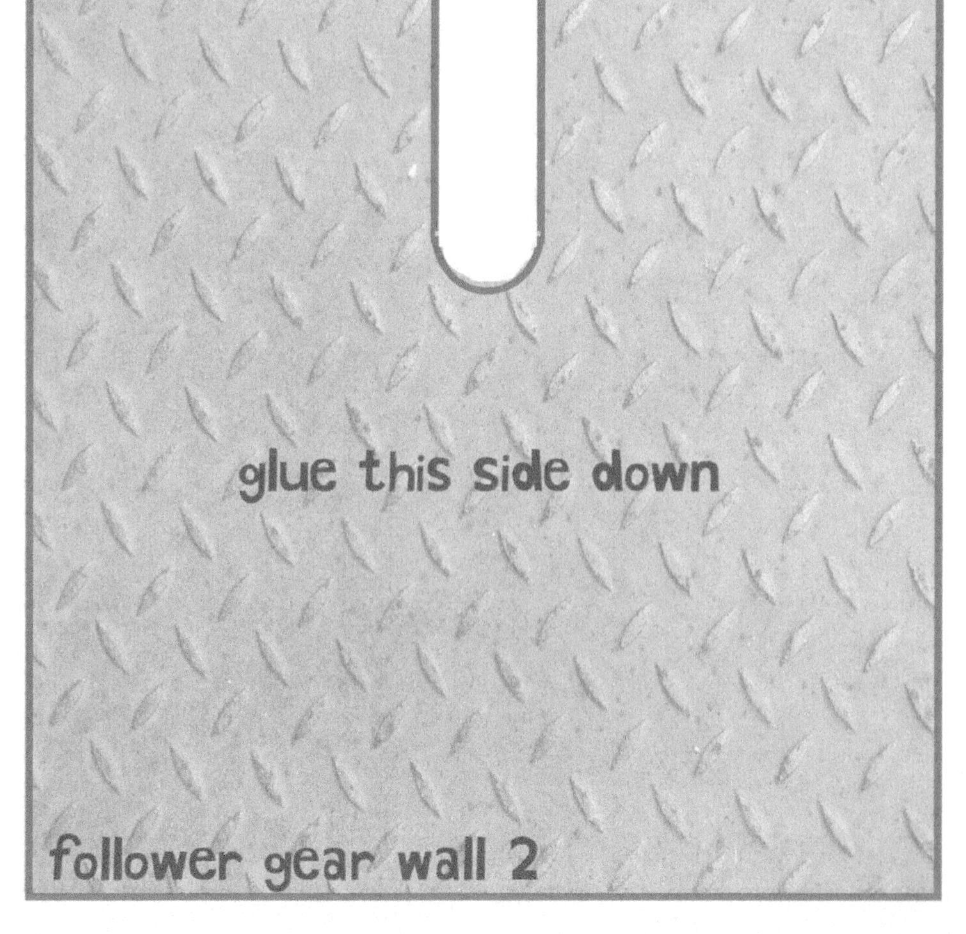
glue this side down
follower gear wall 2

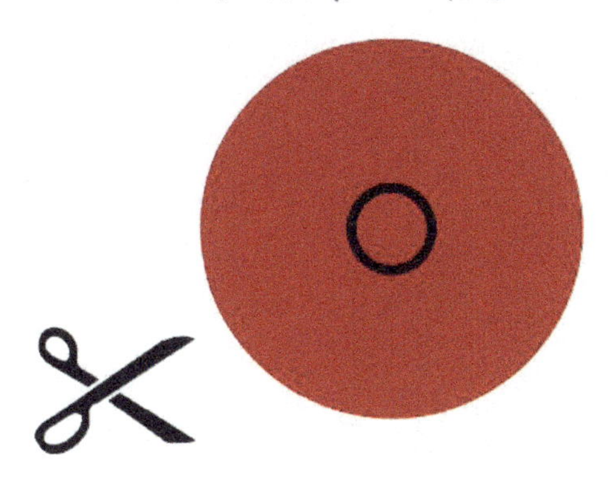

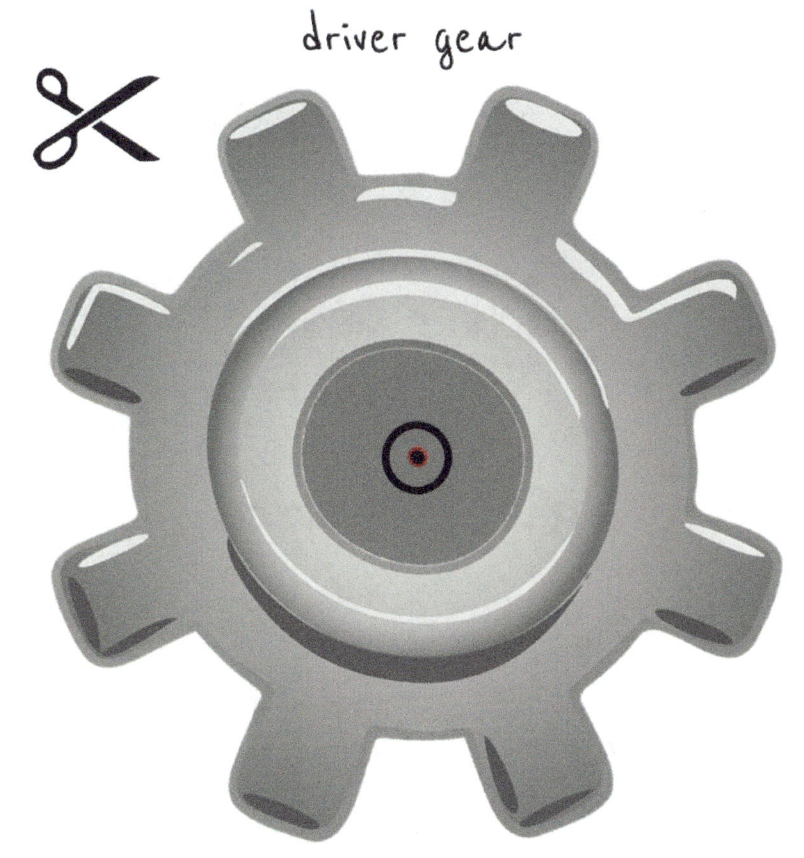

Cut out these templates and glue them to a cardboard sheet. Cut out the cardboard following the shapes. Then poke holes for the pencils to go through.

driver gear
glue this side

glue this side

follower gear
glue this side

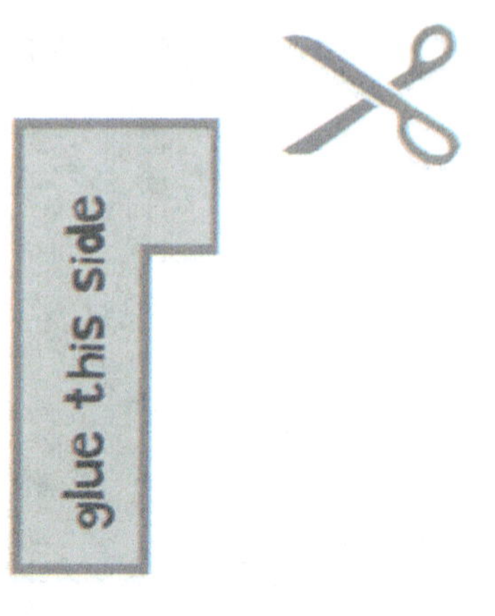
glue this side

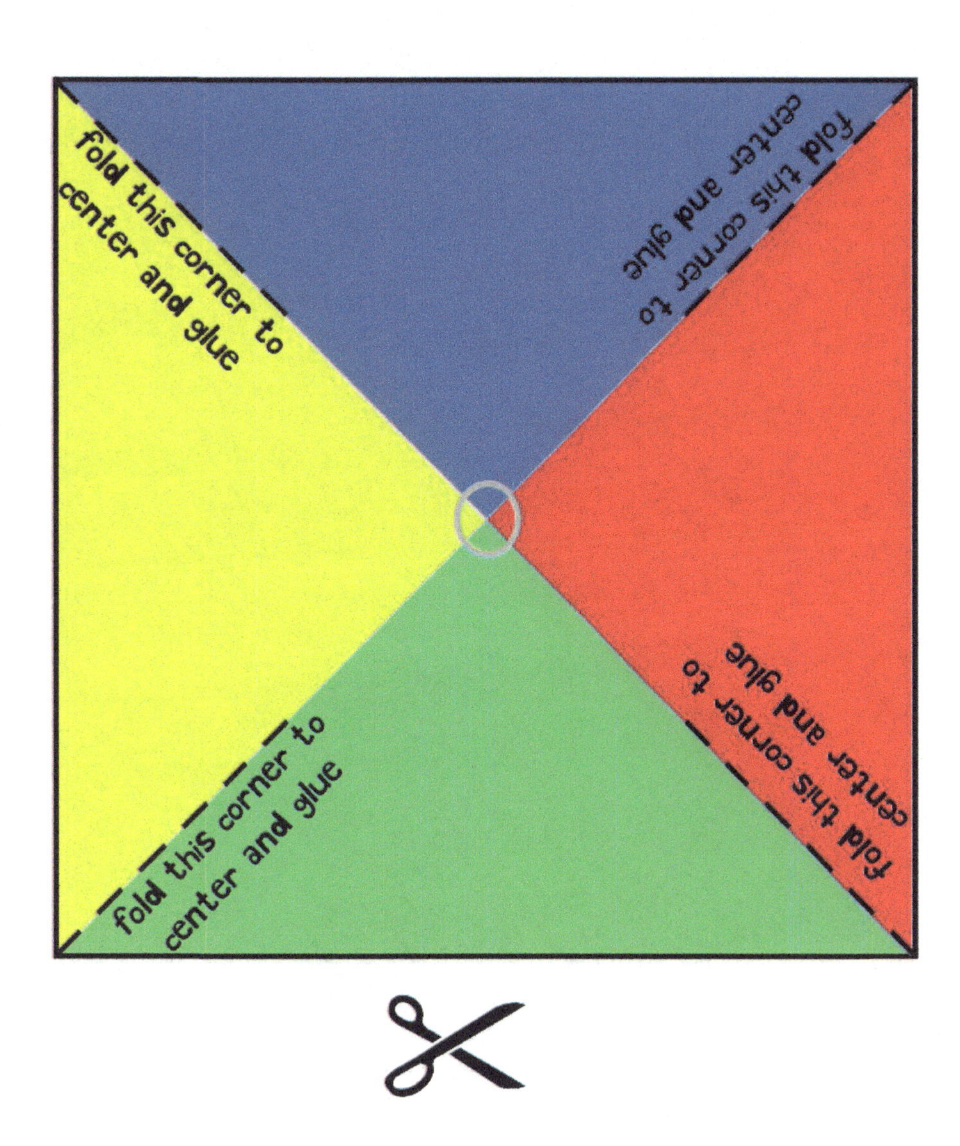

For the PINWHEEL/FAN cut out the square. Then cut on the dotted lines only and stop! (DO NOT CONTINUE CUTTING TO CENTER). Fold the corners to the center and glue.

```
    content: '';
    content: none;
}
table {
    border-collapse: collapse;
    border-spacing: 0;
}
button, input, select, textarea { margin: 0 }
:focus { outline: 0 }
a:link { -webkit-tap-highlight-color: #FF5E99 }
img, video, object, embed {
    max-width: 100%;
    height: auto!important;
```

SOFTWARE ENGINEERING

Learning About Code

00110100111001110

SOLVE

0,1,1,2,3,5,8,

What number comes after 8?

EXPERIMENT 1
BREAK THE BINARY CODE

BINARY ALPHABET

A = 00001
B = 00010
C = 00011
D = 00100
E = 00101
F = 00110
G = 00111
H = 01000
I = 01001
J = 01010
K = 01011
L = 01100
M = 01101
N = 01110
O = 01111
P = 10000
Q = 10001
R = 10010
S = 10011
T = 10100
U = 10101
V = 10110
W = 10111
X = 11000
Y = 11001
Z = 11010

NOTE: FOR UPPERCASE LETTERS ADD 010 IN FRONT, FOR LOWERCASE ADD 011 IN FRONT OF THE BINARY NUMBER.

BINARY CODE IS A LANGUAGE A COMPUTER USES TO PROCESS INFORMATION. LIKE SWITCHES, 1 IS FOR ON, AND 0 IS FOR OFF.

00101 01110 00111 01001 01110

00101 00101 10010 10011

10010 01111 00011 01011

10100 01000 00101

01000 01111 10101 10011 00101

CAN YOU BREAK THE CODE? WRITE THE ANSWER ON THE PAGE ON THE RIGHT.

WRITE YOUR OWN SECRET MESSAGE IN BINARY AND GIVE IT TO SOMEONE AND SEE IF THEY CAN FIGURE IT OUT.

EXPERIMENT 2
LET'S WRITE SOME CODE

What we need!

A device with internet access.

First, let's learn some coding words. Then try to find them in the puzzle.

ALGORITHM Step-by-step action to solve a proble

SYNTAX Rules that make the structure of code.

IDE Integrated development environment. The place where you write the code lines.

DEBUGGING Fixing software problems.

VARIABLE Holds data that can change.

LOOP Repeats a set of instructions.

GIT Manages code changes in a project.

COMPILER Converts code into machine language.

PYTHON An easy coding language.

JAVA A portable programming language.

C Y D S A V C F K R X E G K N

O S B G F A A K U Q U K M U K

M I D H I A V S D A N G E K X

P R U E K X S A B N D Z Y M X

I M F S Y F P U R N H I K Y X

L E Z P H E K Z K I I P K C X

E N G A Q N W T F P A V O J J

R G E T H W J F V G M B E Q H

U W J I N G L A C I L P L R S

J H O T N R R X K T T A M E A

H T W N Y P E Z Z F L Z Y V R

M H X T F Z Q N V T O D A Y H

F D R K P D Y I L C O J A W M

G I D L N I E D S F P H C S L

L M R D O E V E Q Q S A K A Y

E C W B I P G S V A P E M F N

M O P X Y R O K D P F B Y O P

H K K Q D Y I W Z T B Q Z E S

Y T H I D F W E O Q E M Y T W

E R R D N E T Y Z R H V L E L

V I U U I W B M J T U V Y V H

W Q F D F I Y U I S D H T B U

P D T Q K F X R G G D K W X M

F E X Z I J O C A G Z R M S R

L D V I R G I P S V I T X M H

M N A Z L I S Z Y F B N M X C

P N E A C C V E N T S W G M X

H J M Q L X H D T Y H O H T Z

G R W T F K G V A H A O I A S

H W T E V X U E X U U D N Z J

CODE WRITING

Okay ENGINEER let's start with the basics. We have to choose a coding language first. I think PYTHON is pretty easy so let's start with that. Then we need an IDE (the place where we write the lines of code and run the program). You can type in a search engine "IDE for Python" and you will get a bunch of them. I like to use www.online-python.com but it's entirely up to you which one you use. So follow the instructions on the following pages to start writing code.

INSTRUCTIONS

1. GO TO WWW.ONLINE-PYTHON.COM

2. AT THE TOP LEFT CLICK THE " + " SYMBOL.

3. TYPE IN A NEW FILE NAME (KEEP .PY AT THE END)

4. PUT THE CURSOR NEXT TO THE NUMBER " 1 ".

5. TYPE:

PRINT ("I WANT TO BE AN ENGINEER")

ON LINE NUMBER 1.

6. CLICK " RUN "

7. YOU JUST WROTE YOUR FIRST LINE OF CODE.

LET'S TRY AI CODING

1. Go to your favorite AI text website (such as chatgpt).
2. put in the prompt "Create a simple Python program that tells a joke".
3. copy the code.
4. go to www.online-python.com
5. paste the code next to the number " 1 "
6. Click run.
7. study the code. try one of your own.

ENVIRONMENTAL
ENGINEERING
Saving The Planet

SOLVE
$\sqrt{25} =$

EXPERIMENT 1
WATER FROM THIN AIR

What we need!

cardboard

plastic bottle

cutter

screen or fabric that lets air through.

WATER COLLECTOR

DIRECTIONS

1. Cut out strips of the cardboard.
2. Glue the strips into two triangles
3. Cut a plastic water bottle in half.
4. Glue a long strip at the top point of each triangle.
5. Glue the bottle half into the triangle.
6. Glue the mesh/fabric to the cardboard strip at the top so it goes down into the half bottle.
7. Let the contraption sit outside for 24 hours on a hot night.

MATERIALS

1. cardboard
2. water bottle
3. mesh/fabric
4. glue
5. scissors
6. time

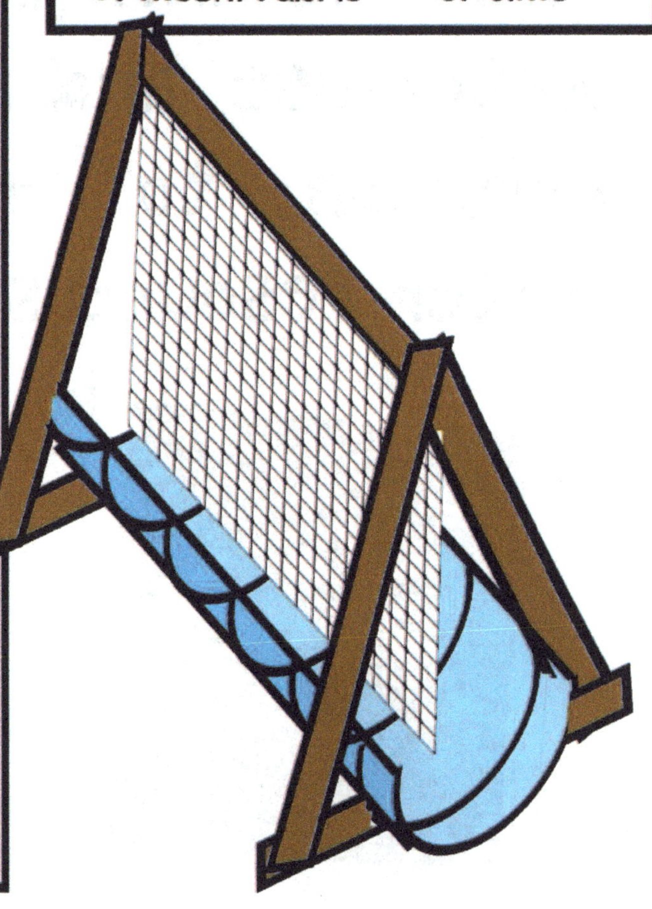

WATER RESOURCE ENGINEER

This is an ENVIRONMENTAL ENGINEER who works to manage and protect the earth's water supply and usage. They develop strategies and technologies for efficient water use, and reduction of water pollution to create sustainable water resources.

EXPERIMENT 2
LET'S MAKE A COMPOSTER

Waste Management Engineers design composting systems to turn waste into valuable resources, reduce landfill size and sustain soil health in small to big-scale projects.

HOW A COMPOSTING TOILET WORKS
THIS TOILET REDUCES WATER WASTE!
Stinky Air Out
BACTERIA
The Good Stuff
on Stinky Air In
Sawdust/Coconut Shavings
The Pee Compartment
The Poop Compartment

COMPOSTER INSTRUCTIONS

1. CUT THE TOP OFF THE PLASTIC BOTTLE.
2. POKE HOLES IN THE BOTTOM OF THE BOTTLE WITH THE TACK.
3. FILL A LITTLE LESS THAN HALF WITH DIRT.
4. ADD PAPER STRIPS + LEAVES.
5. ADD FOOD WASTE.
6. PUT THE BOTTLE TOP BACK ON BUT UPSIDE DOWN.
7. PUT OUTSIDE IN THE SUN.
8. WAIT A WEEK AND CHECK THE RESULTS.

CONGRATS ENGINEER!

DEAR ENGINEER,

You have completed all the engineering tasks in this book. I hope you got some good data and formulated some of your own ideas along the way. Because you worked so brilliantly with all the experiments in this book the esteemed organization called "THE EXCELLENT ECHELON OF ECCENTRIC ENGINEERS" has taken notice of your work and awarded you the most honorable of awards for engineering "THE EFFICACIOUS ELEPHANT EAR" award for ENGINEERING. You can collect your award on the following pages. Thanks ENGINEER, and good luck with your future endeavors!!!

CHEERS

Albert B. Squid

P.S. Oh, I almost forgot. One last thing **ENGINEER.**

SOLVE

$$(8 \div 2 + 15 \times 6 - 4)$$

$$+$$

$$(12 \div 6 \times 2 + .5 - 1)$$

$$+$$

$$(.09 + 5 \times 8 - 4 + 3)$$

$$+$$

$$\sqrt{25}$$

$$=$$

THE EFFICACIOUS ELEPHANT EAR

Is Present To:

____

(your name here)

For outstanding performance in Engineering including but not limited to all disciplines of civil, aerospace, mechanical, software, and environmental engineering areas.

SIGNED BY: Albert B. Squid

LOCATION: ______________________________

Great job Engineer keep up the
amazing work!!!

Albert B. Squid

ABOUT THE AUTHOR

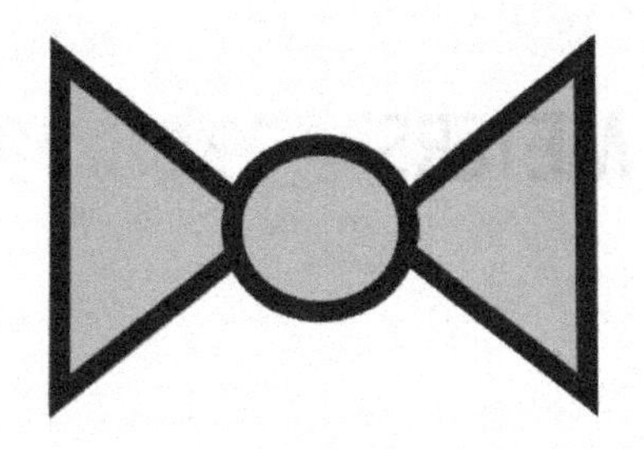

Born to a family of construction peeps, ALBERT B. SQUID was raised on construction sites in Massachusetts. Believe it or not, he holds two degrees in Engineering and Architecture and has worked as an Architect in Boston, Tokyo, and Seoul. In the year 2000, Squid started an independent children's book publishing company in NYC. I had fun doing that.....I mean HE (Albert B. Squid) had fun doing that! After becoming a freelance voice actor, the elusive author's whereabouts are unknown. He was last seen in The Hamptons on the beach talking to a woman with blonde hair about the tide being really high while the man to his left just said,"Qu'est-ce que c'est"

If you have a clue as to where Albert B. Squid might be, let us know at HQ by contacting us at:

nfo@squarerootofsquid.com

NOTE: Although Squid likes to stay out of the public eye, he should be easy to spot with his hat with flaps, mirror sunglasses, and funny bow ties.

ANSWERS

13 + 15 = 28

7 x 9 = 63

39 - 6 = 33

81 ÷ 9 = 9

8 x .3 = 2.4

.1 + .1 = .2

5 x .06 = .3

5 + .06 = 5.06

5 - .06 = 4.94

8981 FEET = 2737.4 METERS

1149 METERS = 3768.72 FEET

5 + 3 x 6 = 23

circumference = 2 x 3.14 x 6

C = 37.7

0,1,1,2,3,5,8,13

binary code: ENGINEERS ROCK THE HOUSE

$\sqrt{25}$ = 5

(8 ÷ 2 + 15 x 6 - 4) = 90

+

(12 ÷ 6 x 2 + .5 - 1) = 3.5

+

(.09 + 5 x 8 - 4 + 3) = 39.09

+

$\sqrt{25}$ = 5

TOTAL= 137.59

YOU THOUGHT THIS BOOK WAS RAD, WE'VE GOT PLENTY MORE
R YOU TO CHECK OUT IN THE "YOU BE THE.........." SERIES!!!

albertbsquid.com

Made in United States
Orlando, FL
13 October 2024